War Department Technical Manual
TM 3-376A

PORTABLE

FLAME THROWER

M2-2

By *War Department* - *16 May 1944*

Washington, D. C.

DISCLAIMER:

This manual is sold for historic research purposes only, as an entertainment. It contains obsolete information and is not intended to be used as part of an actual operation or maintenance training program. No book can substitute for proper training by an authorized instructor.

WAR DEPARTMENT,
Washington, 25, D. C. 16 May 1944

TM 3-376A, Portable Flame Thrower M2-2 is published for the information and guidance of all concerned.

[A. G. 300.7 (21 March 44)]

By order of the Secretary of War:

G. C. MARSHALL,
Chief of Staff.

Official:

J. A. ULIO,
Major General,
The Adjutant General.

Distribution:
R & H (5); Bn 2, 7, 17 (2); C & H 3 (5); IC & H 5 (5); C 2, 7, 17 (2); X. ID: T/O & E 72T, Light Div; 17, Armd Div; IR: T/O 5-192, Hq & Hq Co, Engr Comb Gr; 5-171, Engr Comb Regt; IBn: T/O 5-15, Engr Comb Bn; 5-35, Engr Bn Sep; 5-175, Engr Bn, Comb Regt; 5-215, Armd Engr Bn; 5-475T, Engr Bn, Light Div; IC: T/O 5-16, Hq & Hq & Sv Co, Engr Combat Bn; 5-17, Engr Comb Co; 5-192, Hq & Hq Co, Engr Comb Gr; 5-36, Hq & Hq & Sv Co, Engr Bn (Sep); 5-37, Co, Engr Bn (Sep); 5-176, Hq & Hq Det, Engr Bn, Engr Comb Regt; 5-216, Hq & Hq Co, Armd Engr Bn; 5-217, Co, Armd Engr Bn; 5-476T, Hq & Hq Co, Engr Bn, Light Div; 5477-T, Co, Engr Bn, Light Div.

(For explanation of symbols see Par 26, FM 21-6)

TABLE OF CONTENTS

PART ONE

INTRODUCTION

PART TWO

OPERATING INSTRUCTIONS

INDEX

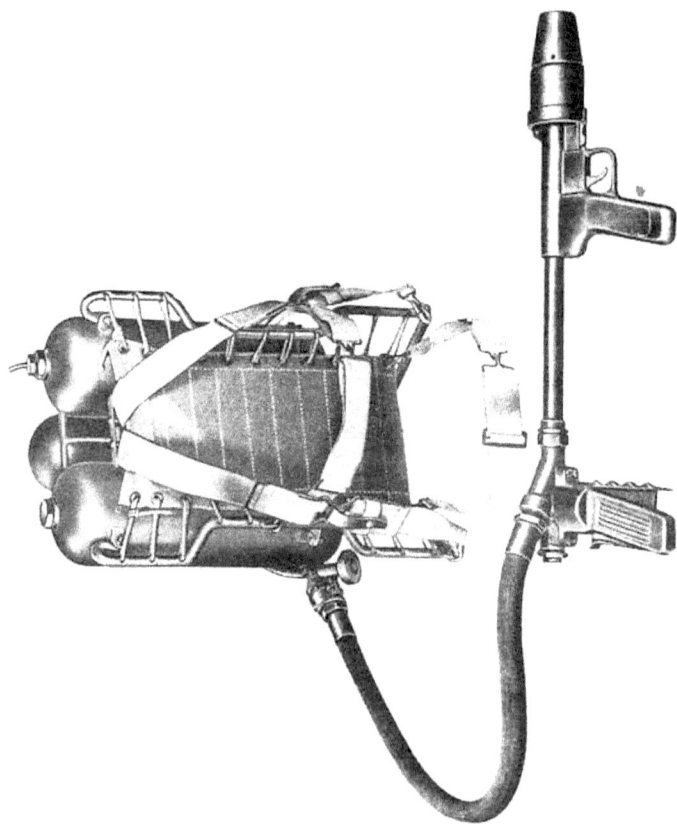

Fig 1. Portable flame thrower M2-2.

PART ONE

INTRODUCTION

Section I GENERAL

1. SCOPE.

a. Arrangement. This manual is published to guide and inform personnel using and maintaining flame thrower, portable, M2-2. Part One contains general information; Part Two is a guide to operation; Part Three gives maintenance procedures. The Appendix discusses shipment and storage procedures, and applicable publications.

b. References. References are listed in the Appendix. The list includes field manuals, technical manuals, and Army Regulations.

2. RECORDS.

Although no standard maintenance forms and records are furnished, an improvised list should be kept of the number of times each flame thrower has been fired. The list indicates when it is necessary to provide the after-six-missions preventive maintenance and lubrication. It should be tacked or glued to the inside surface of the packing-chest lid and each flame thrower should always be returned to its own chest.

Section II DESCRIPTION AND DATA

3. USES OF FLAME THROWERS.

Flame throwers can: a. Penetrate openings, such as embrasures and gun ports, and fill the fortifications with flame and smoke.

b. Burn, asphyxiate, and blind enemy personnel, causing casualties, shock, panic, and abandonment of a fortified position.

c. Ignite combustible parts of shelters and materiel and start detonation of sensitive ammunition and explosives.

Fig 2. Firing with liquid fuel.

Fig 3. Firing with thickened fuel. Thickened fuel has longer range than liquid fuel and burns on target for several minutes.

d. "Shoot around corners," when fuel is fired from dead or blind angles. This is made possible by the billowing and swirling movements of flaming gases. Blazing thickened fuels also ricochet from wall to wall in fortifications.

e. Cause the enemy to close ports, temporarily putting the emplacement out of action and thus protecting the demolition party.

f. Mop up dug-in personnel.

g. Eliminate enemy nests in street or jungle fighting.

4. CHARACTERISTICS AND EMPLOYMENT.

a. Action. Fuel is propelled into the target by a charge of highly compressed air or nitrogen. As fuel leaves the gun of the M2-2 portable flame thrower (Fig 1), it is ignited by contact with flame from charges of incendiary mix held in an expendable ignition cylinder.

b. Bursts. A continuous stream or separate bursts may be fired for approximately 8 to 9 seconds, not including time between the bursts. The five incendiary charges in the ignition cylinder are controlled by the trigger and can ignite several bursts.

c. Range. Portable flame throwers are fired at extremely close or point-blank range for best results. (Par 22) Effective range for liquid fuels (Fig 2) is as far as 20 yards, and for thickened fuels (Fig 3), 40 yards, but underbrush and adverse winds can reduce the distances.

d. Weight. To keep the weight as light as possible and still provide strength to withstand very high pressures, most parts are made of aluminum or sheet steel.

e. Tactics. Two or more flame throwers are generally used on a mission with other weapons of the assault squad. (See FM 31-50, "Attack on a Fortified Position and Combat in Towns."

f. Firers and assistants. One man carries and fires each flame thrower. Well-armed assistants accompany firers to give close protection and to serve as emergency replacements. Whereas the M1A1 portable flame thrower may require the help of an assistant to open the pressure-tank valve, the M2-2 flame thrower pressure-tank valve is located within reach of the firer and is operated by him without assistance. Firers and assistants should be thoroughly trained in operation of the weapon.

g. Charging and filling. In order to replace pressure tanks (cylinders) of earlier types of flame throwers, it is necessary to unscrew and screw threaded connections. Experience has shown that this frequently resulted in damage to threads, leakage, loss of pressure, and loss of range. It is also necessary to use tools to replace each pressure tank (cylinder). The design of the M2-2 flame thrower eliminates these difficulties. The tank group (Fig 4) may be charged and filled as a unit with or without gun and hose. The quick-connecting tank coupling permits rapid interchanging of empty and full tank groups by the firers or as-

FUEL
TANK

PRESSURE
TANK

FUEL
TANK

PRESSURE-
TANK VALVE

TANK
CONNECTOR

TANK
COUPLING

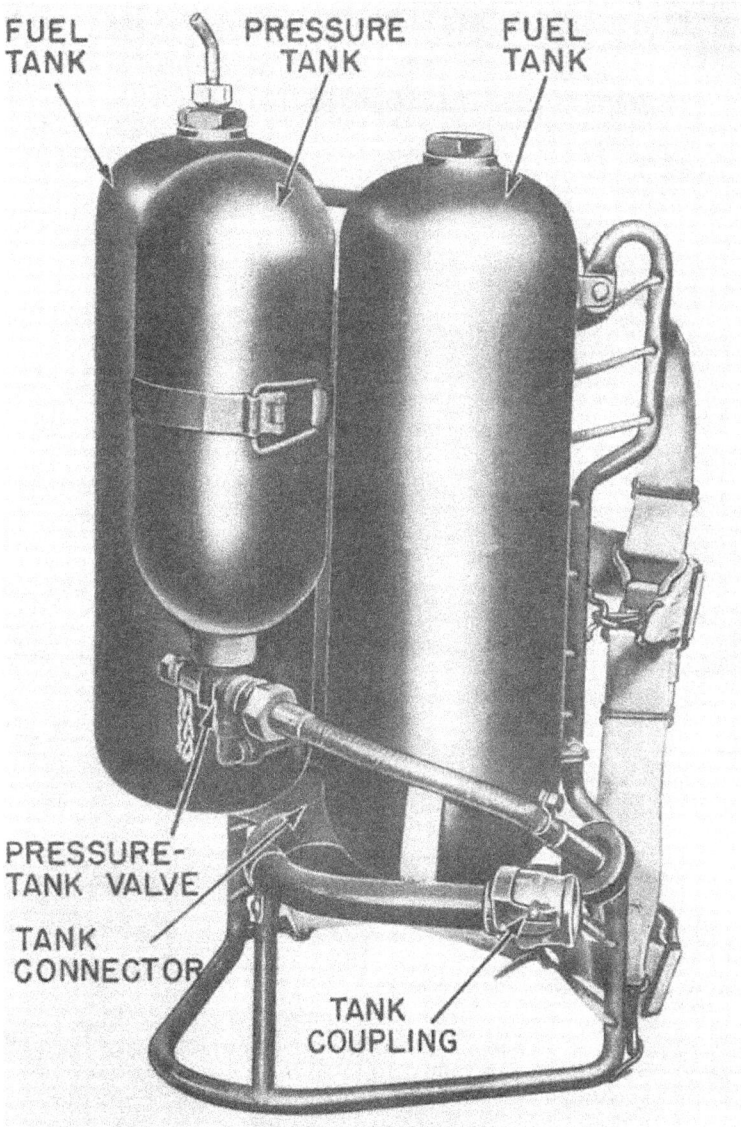

Fig 4. Tank group.

sistants. This is done without tools, takes very little time, and cannot cause leakage, loss of pressure, and loss of range due to damaged threads.

5. DESCRIPTION AND FUNCTIONING.

The flame thrower consists of two major groups: tank group and gun group. Detailed descriptions of assemblies and parts are included in Paragraphs 66 through 76.

a. Tank group. (Figs 4 and 5) Carried upon the firer's back, the tank group holds fuel and pressure. The tank group may be identified as tank, fuel, portable flame thrower, M2, assembly D81-1-482. It consists principally of:

(1) Two fuel tanks, holding a total of 4 gallons of fuel, and joined by a tank connector to form a single fuel reservoir.

(2) Pressure tank, charged with highly compressed air or nitrogen used to propel fuel from the fuel tanks through the gun to the target. The tank is large in capacity to assure ample pressure and uniformly long range throughout the firing.

(3) Pressure-tank valve, which releases air or nitrogen through the pressure regulator to the fuel tanks. The valve can be opened by the firer without the assistance required in the case of the M1A1 flame thrower.

(4) Pressure regulator, which automatically assures delivery of air or nitrogen to the fuel tanks at the proper pressure. The regulator is located in a position where it cannot easily be damaged.

(5) Carrier, which supports the tank group on the firer's back and shoulders and secures it to his body. It includes body and shoulder straps and quick-releasing fasteners.

b. Gun group. (Fig 6) Carried, aimed, and operated by the hands of the firer, the gun group ignites the fuel and directs the flame into the target. It includes:

(1) Fuel hose, which conveys fuel from the tank group to the gun. The fuel hose may be requisitioned as hose, fuel, portable flame thrower, M1, assembly B81-1-498.

(2) Gun, which ignites the fuel and directs it to the target. The gun may be identified as gun, portable flame thrower, M2, assembly D81-1-405. It consists of:

(a) Fuel valve, which discharges fuel through the barrel. The valve is operated by squeezing the valve lever and the grip safety, which are on opposite sides of the valve grip. The valve also includes a barrel from which the fuel is ejected. The ignition head is supported on the front of the barrel.

(b) Ignition head, which ignites the fuel as it passes from the nozzle of the barrel. With each pull of the trigger on the front grip, one of five charges of incendiary mix in an ignition cylinder is ignited. This pilot flame ignites the fuel as it is propelled from the gun.

Fig 5. Tank group with carrier folded back to show construction.

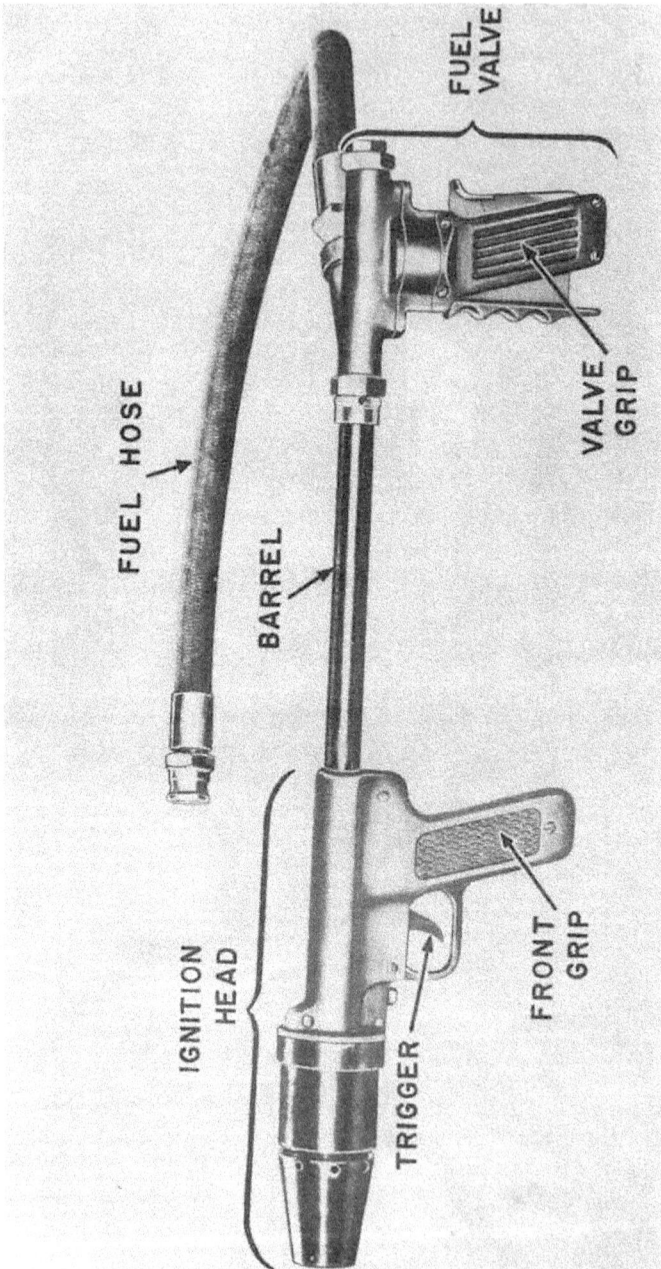

Fig 6. Gun group of portable flame thrower M2-2.

6. IDENTIFICATION INFORMATION.

The words "Chemical Warfare Service," model numbers, serial numbers, lot numbers, weight, cubage, manufacturers' names, contract number, and date of packing are indicated on the packing chest or the equipment. The numbers and letters shown on the equipment should be referred to when repairs are required. The tank group and the gun (without the fuel hose) may each be marked "M2" and the fuel hose may be marked "M1," although all of these are components of the M2-2 portable flame thrower.

7. DIFFERENCES IN MODELS.

a. M2-2 and E3 portable flame throwers. Portable flame thrower M2-2 is identical in all important respects with portable flame thrower E3. (The E3 flame thrower, when standardized with some modifications, became the M2-2.) Operation and maintenance of the M2-2 and E3 are in general the same, and the parts are interchangeable.

b. M2-2, M1, and M1A1 portable flame throwers. Portable flame thrower M2-2 has the same fuel capacity but differs in construction from portable flame throwers M1 and M1A1. Parts are not interchangeable except as stated in Paragraph 8.

8. INTERCHANGING PARTS WITH M1 OR M1A1 FLAME THROWER.

To use an M2-2 gun with tank group (fuel unit) of an M1 or M1A1 portable flame thrower:

a. Remove fuel hose from M2-2 gun.

b. Screw a 3/4-inch by 1/2-inch pipe bushing into the side opening of the fuel-valve body. This bushing is furnished in the spare parts kit of each M2-2 portable flame thrower. (Par 10)

c. Screw the fuel-hose assembly of the M1 or M1A1 flame thrower into the 1/2-inch opening of the bushing, using a wrench to make a tight connection.

9. DATA.

All data are approximate.

a. Range. See Paragraph 21.

b. Duration of fire.

(1) Fuel.

(a) Continuous discharge of approximately 8 to 9 seconds, or

(b) Several short bursts totalling approximately 8 to 9 seconds (not including time between bursts).

(2) Ignition cylinder. Five charges in each cylinder, 8 to 12 seconds per charge.

   <u>c</u>. Weights.

	Pounds
Portable flame thrower M2-2, empty, in shipping chest (including the chest and all contents).	110
Portable flame thrower M2-2, empty	43
Portable flame thrower M2-2, filled with fuel	68 to 72
Tank group, empty	35
Tank group, filled with fuel	60 to 64
Gun group	8

   <u>d</u>. Dimensions.

	Inches
Gun, length	30
Fuel hose, length	37
Tank group, height	27
Tank group, width	20
Tank group, breadth	11
Packing chest	34 x 23 x 19

    (Cube of packing chest: 8-1/2 cubic feet)

   <u>e</u>. Capacity of weapon.

Ignition cylinder (M1 or E1). . . .1 (which includes 5 incendiary charges)

Fuel. . . . . .4 gallons plus void for air or nitrogen

   <u>f</u>. Pressures.

	Pounds per sq. in.
Pressure tank	1,700 to 2,100
Fuel tanks	350

   <u>g</u>. Ratio of expended supplies. For every 100 complete fillings of the flame thrower, the following supplies are normally expended:

   (1) Nitrogen contained in fifteen 220-cubic-foot cylinders or an equivalent volume of compressed air. (Eleven cylinders are expended if the four-place arrangement described in Paragraph 32 is used.)

   (2) 450 gallons of fuel (400 gallons plus 50 gallons for spillage, spoilage, and evaporation).

   (3) 100 ignition cylinders.

   (4) If thickened fuel is used, 135 pounds (in cans of 5-1/4 pounds each) of U. S. Army fuel thickener.

Section III TOOLS,
PARTS, AND ACCESSORIES

10. ITEMS WITH EACH FLAME THROWER.

The items listed below or their equivalents (Fig 7) are included in each M2-2 flame-thrower packing chest, in addition to the flame thrower. Numbers listed with items are Chemical Warfare Service stock numbers.

a. Kit, tool, for portable flame thrower M2-2, assembly B81-6-50.

b. Kit, spare parts, for portable flame thrower M2-2, assembly B81-6-52.

c. Cylinder, ignition, portable flame thrower M1. (6 cylinders, in 3 cans containing 2 each)

d. Technical Manual 3-376A, "Portable Flame Thrower M2-2."

e. Gun mounting board. (Fig 10)

f. Plug, coupling, E81-1-514 (for use in tank coupling when filling tank group with gun detached).

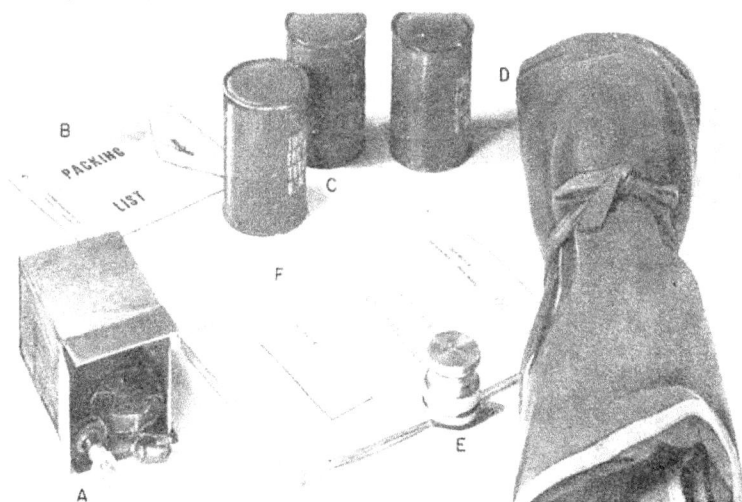

Fig 7. Items packed in chest with each flame thrower: A — Spare parts kit; B — Packing list; C — Three cans of ignition cylinders; D — Tool kit; E — Coupling plug; F — TM 3-376A, "Portable Flame Thrower M2-2."

11

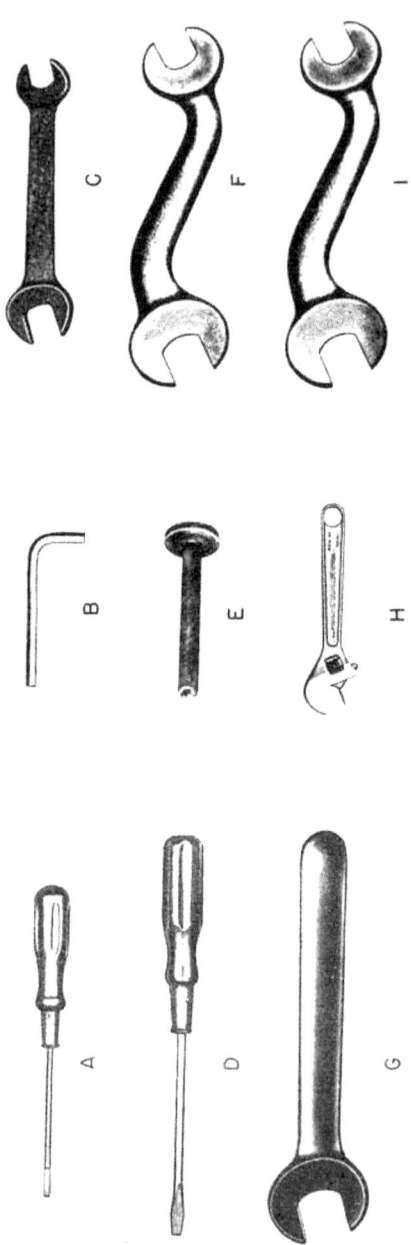

Fig 8. Contents of tool kit:

A. 1 Screw driver, cabinet, 4-1/2-inch blade length, 3/16-inch blade diameter, H22-50-13.
B. 1 Wrench, hex, 1/8-inch across flats for 1/4-inch socket-head set screws, H22-49-12.
C. 1 Wrench, engineers', double head, 3/4-inch and 7/8-inch openings, 9 inches approx length, H22-49-115.
D. 1 Screw driver, common, 6-inch blade length, 5/16-inch blade diameter, H22-50-6.
E. 1 Wrench, valve-adjusting, assembly, A81-6-48.
F. 1 Wrench, heavy "S", 1-3/8-inch and 1-1/2-inch openings, 12 inches approx length, H22-49-113.
G. 1 Wrench, engineers', single head, 1-1/8 inch opening, 10-1/2 inches approx length, H22-49-31.
H. 1 Wrench, adjustable, single end, 6 inches approx length (crescent type), H22-49-67.
I. 1 Wrench, heavy "S", 1-3/8-inch and 1-3/4-inch openings, 12 inches approx length, A81-6-49.

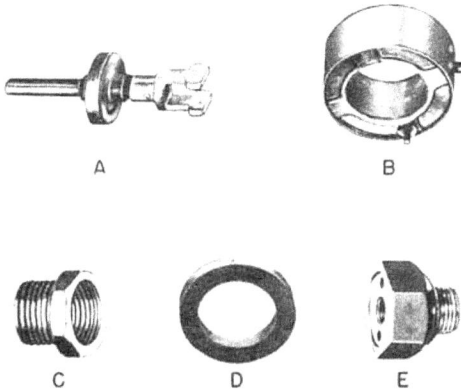

Fig 9. Contents of spare parts kit:
  A. 1 Diaphragm, valve, assembly, A81-1-416.
  B. 1 Case, spring, assembly, B81-1-444.
  C. 1 Bushing, pipe, head, 3/4-inch by 1/2-inch
      (galvanized iron), H98-5-93.
  D. 2 Washers, coupling, A81-1-513.
  E. 3 Heads, safety, R81-1-561.

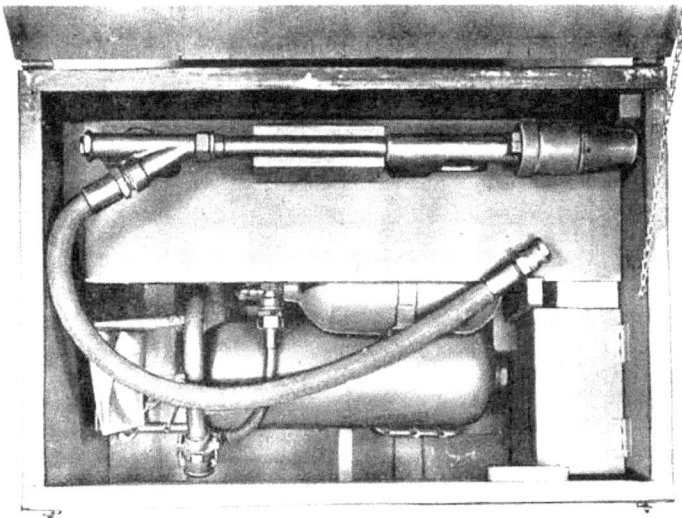

Fig 10. Packing chest open, with gun on mounting board. Tool kit,
spare parts kit, and cans of ignition cylinders in boxes at right.

PART TWO

OPERATING INSTRUCTIONS

Section IV GENERAL

11. SCOPE.

Part Two of this manual is for the guidance of operating personnel. It includes information on the controls and on operation.

Section V SERVICE UPON RECEIPT OF EQUIPMENT

12. NEW EQUIPMENT.

Upon receipt of a new flame thrower, the following procedure should be carried out:

a. Cut packing-chest steel straps and seals with pliers.

b. Remove the screws, if present, from top of chest.

c. Open two latches at front of chest.

d. Lift lid backward and connect chain from inside of chest to inside of lid.

e. Remove moistureproof paper.

f. Remove gun from carton. After removing waterproofing tape from ends of hose, connect hose and gun. (Par 17)

g. Remove mounting board and place gun with hose on the board as shown in Figure 10.

Fig 11. Screwing deflector tube in safety head on left fuel tank.

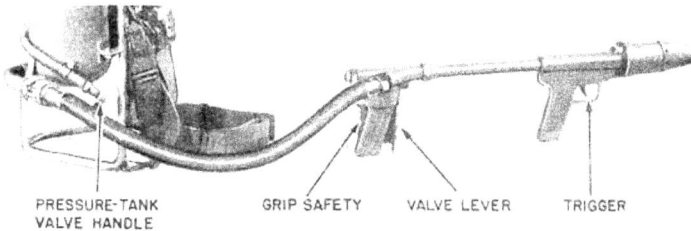

Fig 12. Controls for operation of portable flame thrower M2-2.

h. Remove spare parts kit, tool kit, cans of ignition cylinders, and other items from packing chest.

i. Compare contents with packing list found in or on packing chest. Inspect all contents carefully for completeness, correct adjustment, and good condition.

j. Insert deflector tube in safety head on left fuel tank. (Fig 11) Outlet should face to rear and at a 45-degree angle to operator's left shoulder. (Fig 18) Screw in deflector tube by hand; do not use wrench on deflector tube. Tighten lock nut with wrench.

k. Before use on a mission, test-fire the weapon. (Par 56 b)

l. Save the packing chest for storage of the equipment when flame thrower is not being carried on a firing mission or serviced.

13. USED EQUIPMENT.

When they apply, the same steps should be taken as in Paragraph 12. Any worn or damaged parts should be replaced. Areas where paint has worn off should be touched up with fresh paint.

Section VI   CONTROLS

14. CONTROLS.

The firer uses the pressure-tank valve handle, the trigger, and the valve lever and grip safety (Fig 12) in succession as follows:

a. Valve handle. The pressure-tank valve is operated by turning a handle on the valve flexible shaft within reach of the firer. Counterclockwise operation of handle releases pressure to the fuel tanks. Clockwise turning closes the valve.

b. Trigger. The trigger is at the front grip of the gun. Pulling the trigger vigorously ignites an incendiary charge in the ignition cylinder. This in turn ignites the fuel as it leaves the gun. The trigger action also causes the ignition cylinder to revolve one-fifth of a turn, presenting another charge for firing. Each of the five charges may thus be used in rapid succession, if necessary, by pulling the trigger vigorously as many as five times.

c. Valve lever and grip safety. The valve lever and grip safety are mounted on opposite sides of the valve grip of the gun. When both controls are compressed, fuel is propelled from the gun. If either the valve lever or the grip safety is not compressed, the fuel valve remains closed and the fuel remains in the weapon.

Section VII OPERATION UNDER
USUAL CONDITIONS

15. TRAINING.
Effective use of the M2-2 portable flame thrower can be achieved only by diligent practice with the weapon. Untrained firers or assistants should never be sent on a mission.

a. Practice. Firers should practice under varying conditions of wind, range, elevation, depression, and traverse. The shortness of the total firing time (approximately 8 to 9 seconds) demands split-second judgment and coordination.

b. Use of water in training. Water may be used (instead of fuel) for elementary practice firing. Ignition cylinders are not used with water. The water under pressure may cause serious injuries to personnel at 10 yards. After practice with water, the gun should be disassembled (Pars 73 through 76), cleaned and dried piece by piece, and lubricated. (Par 49)

c. Use of fuel in training. When using fuel in training, select or prepare a practice field of fire which provides at least 125 yards for range and 30 yards for spread. If the field contains dry grass, brush, or other flammable material, a fire-fighting squad should be available with equipment and source of water. Assistants and observers should stay well behind the firer because of danger from wind shifts. See Paragraph 40 for additional precautions.

16. CHARGING, FILLING, AND SERVICING.
Before use on missions or for training, flame throwers must be charged, filled, and serviced. Charging with compressed air or compressed nitrogen is described in Paragraphs 32 and 33; filling with fuel in Paragraphs 34 through 40; and servicing in Paragraphs 50 through 56. Test for pressure. (Par 53 d)

17. CONNECTING TANK GROUP AND GUN GROUP.
If a charged and filled tank group has been brought up to replace an emptied one:

a. Place the new tank group on the ground with the tank coupling on top. If the filling is thickened fuel, allow the tank group to rest in this position for from 1 to 2 minutes.

b. Remove coupling plug from new tank group and disconnect gun group from emptied tank group. Place unthreaded end of fuel hose in tank coupling and lock in place. (Par 70)

c. Lock the coupling plug in the emptied tank group.

18. LOADING WITH IGNITION CYLINDER.

a. General. Just before the start of a mission, load an unused ignition cylinder into the ignition head. (M1 and E1 ignition cylinders are identical and may be used interchangeably.) Cylinders are packed two to a can. Do not open cans until ready to load for a mission. The second cylinder in the can should be used in another flame thrower on the same mission or as soon as possible after opening the can. Partly used cylinders may be employed in training.

b. Precautions. Care must be taken, whenever cylinders are handled, to avoid any blows or pressure against the metal match ends. (Fig 13) Face, hands, and other parts of the body should never be exposed to front of cylinder or front of gun.

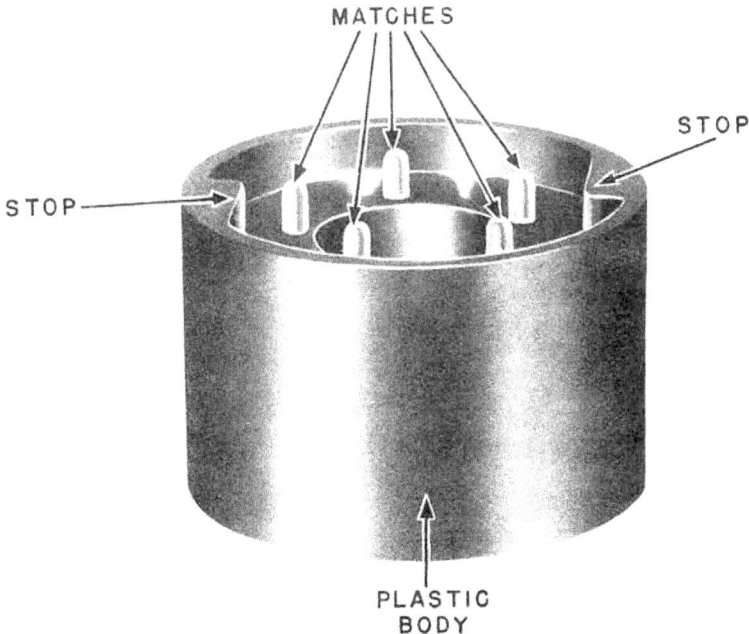

Fig 13. Ignition cylinder before use.

17

c. Procedure. Loading procedure is as follows:

(1) Unscrew and remove ignition shield. (Fig 14)

(2) Place ignition cylinder on end of barrel (Fig 15), being careful not to grasp cylinder by its ends.

(3) Raise nozzle end of gun so cylinder slides down against the spring case of the ignition head. (Fig 16) If necessary, rotate cylinder so it slips down all the way. Do not force cylinder into place as forcing may prematurely ignite it.

(4) Rotate spring case and ignition cylinder clockwise as far as they turn freely.

(5) Place ignition shield over cylinder. Engage the slot in the shield on the spring-case pin.

(6) Turn shield, screwing it onto ignition-head body. Make sure the threads engage during the first turn of the shield. When the slot on the shield engages the latch on the ignition head (Fig 17), the gun is loaded.

(7) If shield cannot be turned by hand tight enough to engage latch, unscrew shield. Then turn shield backwards until threads engage and repeat (6) above.

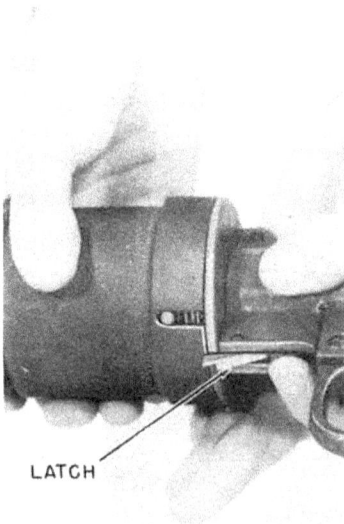

LATCH

Fig 14. Unscrewing ignition shield, with pressure on latch.

Fig 15. Placing ignition cylinder on gun. Care must be taken to avoid striking or pushing metal matches of cylinder.

Fig 16. Ignition cylinder in place on gun before replacement of
ignition shield.

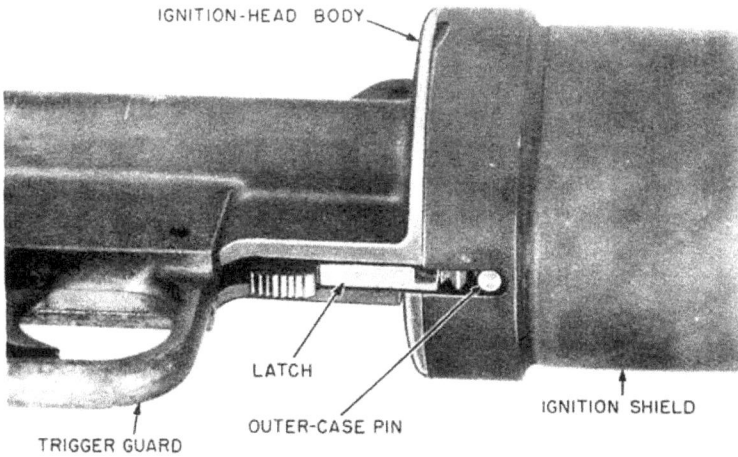

IGNITION-HEAD BODY

LATCH

OUTER-CASE PIN

TRIGGER GUARD

IGNITION SHIELD

Fig 17. Ignition head assembled for firing of gun.

19

Fig 18. Tank group adjusted on firer.

19. CARRYING THE TANK GROUP.

The tanks are supported on the firer's back and secured to it by two shoulder straps and two pairs of body straps. (Fig 18) The straps may be adjusted by the buckles to fit the operator. The shoulder straps pass over the shoulders and under the arm pits; the lower body straps are clasped tightly in front of the body; and the upper body straps are clasped across the chest to prevent the shoulder straps from slipping and the tank group from rolling off the back. Adjustments to the various straps should be made until the unit is carried with the bottom of the fuel tanks at the small of the operator's back. The tank group should fit snugly so that it does not shift if the operator changes position quickly.

Fig 19. Carrying the gun, with hands in position to fire.

20. CARRYING THE GUN.

The procedure for carrying the gun is as follows:

a. Carry the gun with the hose at the right side. (Fig 19)

b. Grasp the valve grip with the right hand and the front grip with the left hand, being careful not to operate the controls until ready to fire.

c. Keep the gun pointed away from friendly personnel at all times.

d. Do not face the front of the gun at any time. Even when no fuel is being ejected, the incendiary charges of the ignition cylinder can cause severe burns.

e. Keep the gun dry and clean if possible. Avoid getting dirt or foreign matter into the weapon.

f. Avoid rough handling.

g. Wear gloves if available.

h. Carry any extra ignition cylinders only in metal containers.

21. OPENING PRESSURE-TANK VALVE.

The release of pressure into the fuel tank causes a hissing sound. Therefore, open the pressure-tank valve while still out of hearing range of the enemy. Do not, however, open it prematurely because of the possibility of pressure leaks. To prevent frothing of the fuel, keep the tank group in as nearly an upright position as possible when opening pressure-tank valve. Be sure to turn the valve handle all the way in a counterclockwise direction. Stiffening of the fuel hose occurs when the pressure-tank valve is opened.

22. RANGES.

Firers and assistants should learn to judge ranges by frequent practice under varying conditions. The firers should be trained to approach as close as practicable to the target and to fire if possible at point-blank range for the greatest results.

a. Point-blank range. (1) Effects. At very close (point-blank) range almost all of the burning fuel can be fired at great velocity directly through ports and openings into the target. Maximum casualties and damage are caused in the hostile position.

(2) Protection. Common sense precautions are taken to prevent casualties to friendly personnel from possible ricochet or rebounding of flame. If the target includes a vertical wall at a right angle to the firer or other friendly personnel, the weapon should not be fired at closer than 7 to 10 yards. When the weapon is fired at small openings in a bunker or pillbox, the firer and other members of the assault squad should not approach closer than 7 to 10 yards from the target.

b. Other effective ranges. (1) Open fields of fire. When thickened gasoline is used, portable flame throwers may fire with considerable effect as far as 40 yards under normal conditions, depending on wind direction and wind speed. Under the same conditions, liquid fuel may be effective at 20 yards. Results and accuracy are not as great as at point-blank range.

(2) Jungle or thick underbrush. If the target is located in jungle or thick underbrush without cleared fields of fire, the effective range of the flame thrower is reduced by as much as one half, depending on the nature and density of the vegetation.

c. Ineffective ranges. Although the flame may reach considerably farther than the ranges stated in b (1) above it may be useless because of the steep angle of descent and because much of the fuel is burned before it reaches the target.

23. WIND DEFLECTION.

Wind is an important factor because of the low velocity of the flaming fuel. Wind can lengthen, shorten, or deflect the flame.

a. Head winds. Head winds of more than 5 miles per hour tend to carry heat or even flame back toward the firer. Liquid fuel should not be fired into a head wind of more than 5 miles per hour. The range and accuracy of thickened fuels is reduced.

b. Following winds or very light winds. Best results are obtained under these conditions.

c. Cross winds. When firing at or near maximum range, cross winds deflect, break up, and disperse the flame. They also reduce the range.

24. FIRING POSITIONS.

a. Ease of aiming. The flame thrower can be fired from any position that permits sufficient freedom to aim the weapon, subject to the conditions in b, c, and d, below. This includes standing, kneeling, and prone. In some instances, flame throwers have been fired with tank groups resting on the ground or on skids. If used in this way, the tops of the fuel tanks must be propped up to conform to b, below.

b. Angles of the tanks. When firing, the bottoms of the fuel tanks must always be substantially lower than the tops. The tops of both tanks must also each be the same distance above the horizontal and neither tank should be tilted to one side. Otherwise, only a small part of the fuel may be blown from the tanks.

c. Recoil. Stability must be sufficient to withstand the recoil from the gun. If possible, the firer should hold the gun snugly against his right side to support it and to absorb its recoil.

d. Protection. Full advantage should be taken of cover and concealment, such as shell craters and vegetation.

25. AIMING.

a. Sighting. There are no sights on the gun because of the short range from which it is fired, the variety of fuels used, and the marked effects of wind. (Par 23)

b. Fortifications. When firing at a fortified position, flame must be directed into openings (gun ports, firing slits, ventilation screens, doorways). Flame inside gives the desired effects, but flame on the outside has little effect on personnel within.

c. Thickened fuel. (Figs 3 and 20) When firing at or near maximum range, it may take several seconds for a burst of thickened fuel to carry through the air to the target area. Short bursts may result in misses at long range for this reason. Skill in aiming is particularly important with thickened fuel.

d. Liquid fuel. With liquid fuel, the greatest effect may be obtained by placing the flame directly on the target. (Fig 21)

26. FIRING.

With pressure-tank valve open:

a. Pull trigger. Pull the trigger rapidly and vigorously. A flash should appear at the front of the gun. This shows that an incendiary charge of the ignition cylinder has been ignited. Release the trigger. (If the flash does not appear, pull the trigger

Fig 20. Thickened fuel flame hitting and clinging to target. Fuel burns for several minutes.

Fig 21. Flame (liquid fuel) hitting target.

again, or as often as necessary up to five times, until a flash appears.)

b. Squeeze fuel valve. Immediately after pulling trigger, compress the valve lever and grip safety vigorously with the right hand. Burning fuel will be propelled from the gun.

c. Adjust fire. Direct the flaming fuel at the target. Continue to squeeze the valve lever and grip safety throughout the burst. When thickened fuel is fired, follow the fuel with eyes to the side of the stream in order to observe and correct aim. (If eyes are directly behind the stream, the flame may obscure the target.)

27. CEASING OR INTERRUPTING FIRE.
To cease or interrupt firing, release the controls.

28. ADDITIONAL BURSTS.
To fire additional bursts, repeat procedure followed in Paragraphs 26 and 27, keeping in mind that there are five incendiary charges in the ignition cylinder and that the total firing time, not including time between bursts, is approximately 8 to 9 seconds. Each of the five incendiary charges in the ignition cylinder burns for from 8 to 12 seconds.

29. SOAKING THE TARGET.
When liquid fuel is used, it may be desirable to soak the target with fuel first and ignite it afterward. To do this, fire one or two short bursts without pulling the trigger. Then follow with an ignited burst, as in Paragraph 26.

30. AFTER FIRING.
When the firer has returned from his mission, he should:

a. Remove and discard the ignition cylinder, as follows:

(1) Point gun at the ground.

(2) Press latch. (Fig 14)

(3) Unscrew the ignition shield and allow ignition cylinder to fall out. (Be careful to keep the hands away from the front of the cylinder.)

(4) Save the partly used cylinder for training use or destroy it by firing from gun after fuel tanks have been emptied. For information on care, handling, and storage of cylinders, see Paragraph 31.

b. Close the pressure-tank valve by turning valve handle clockwise (to conserve remaining pressure in pressure tank).

c. Point the gun away from personnel and blow out the remaining fuel, if any, from the fuel tanks by squeezing the valve lever and grip safety until there is no further discharge. The trigger should not be used during this operation.

d. Take off tank group from the back.

e. Inspect, clean, and maintain the flame thrower (Pars 55 and

56) or, if experienced maintenance personnel is close at hand, turn the weapon over to them for servicing.

f. After servicing, place the weapon in the packing chest (Par 77) for protected storage, or prepare it for the next mission. (Pars 50 through 53)

Section VIII AUXILIARY EQUIPMENT

31. IGNITION CYLINDER.

a. Description and functioning. (Figs 13 and 22) Either the M1 or E1 ignition cylinder may be used. It fits over the fore part of the barrel assembly and is revolved by the spring case. (Par 76) The five incendiary charges in the cylinder are spaced sufficiently far apart in the plastic body to prevent their igniting one another. Lead-foil seals, plastic closure plates, and waterproof cement make the unit comparatively waterproof.

b. Action. When the trigger rod is pushed forward, one of five metal matches tipped with red phosphorus scratches an igniting mixture. The ignition carries to a starter mix and to a few grains of black powder on top of the incendiary charge. The black powder blows the foil seal and closure plate clear of the flame thrower,

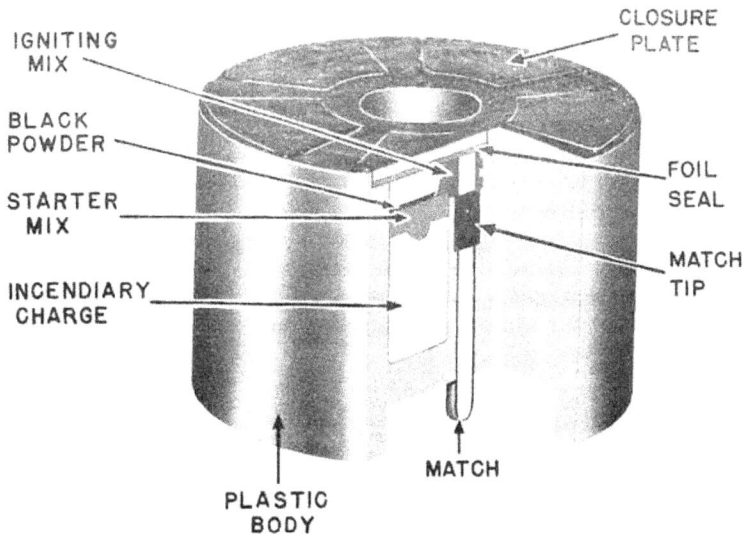

Fig 22. Cutaway view of ignition cylinder (M1 or E1).

and the incendiary charge ignites the fuel as it is discharged from the nozzle. The incendiary charge burns for from 8 to 12 seconds.

c. Packing. Ignition cylinders are packed two per waterproof can. Three cans are furnished with each flame thrower. Fifty cans (100 ignition cylinders) are contained in each packing box of extra cylinders.

d. Care, handling, and storage. Ignition cylinders contain hazardous incendiary material and must be handled with due care. The following precautions should be observed.

(1) Opening cans. Do not open cans containing cylinders until ready to load for a mission. (Par 18) If an extra cylinder remains in an opened can, use it as soon as possible. Any defective cylinders, such as those with damaged closure plates, should be destroyed. (Pars 30 and 46) Moisture may affect the cylinders and all possible care should be taken to avoid exposing them to dampness.

(2) Handling cylinders. Pressure on any of the five metal matches (Fig 13) may ignite an incendiary charge in the cylinder. Care must be exercised to avoid putting pressure on the projecting ends of the matches except when firing the weapon. Ignition cylinders and cylinder containers should be protected against shock. Boxes and cans containing cylinders must not be thrown or dropped.

(3) Storing containers. Containers of ignition cylinders are best stored in a dry, well-ventilated place, out of the direct rays of the sun, well protected against excessive temperatures. Smoking is not permitted and matches are not used where ignition cylinders are stored.

32. CHARGING PRESSURE TANK.

a. General. The pressure tank of the flame thrower must be fully charged with compressed air or compressed nitrogen before the start of a mission. For the M2-2 flame thrower, a pressure of at least 1,700 pounds per square inch is required. This may be provided either by the use of an air compressor capable of producing a pressure of at least 1,700 pounds per square inch, or by the use of commercial cylinders. The filling and charging lines from the service kit are used in conjunction with the cylinders. Before and after charging, follow the procedures described in Paragraphs 51 and 55.

b. Charging from air compressor. Compressor, air, gasoline engine driven, 7CFM, M1, is a self-contained, skid-mounted machine designed for use with flame throwers. It is capable of charging pressure tanks of flame throwers and large 200- or 220-cubic-foot commercial cylinders as well. Instructions for use of the compressor will be found in the manual accompanying it.

c. Charging from cylinders. If an air compressor is not available, it is necessary to use cylinders containing nitrogen or air.

(1) Volume and pressure. Cylinders come charged with 200 to 220 cubic feet of air or nitrogen. Since cylinders with 220 cubic feet of air or nitrogen have a higher initial pressure, it is recommended that they be procured, if obtainable. All cylinders used must have a pressure of at least 600 pounds per square inch. One or more of the cylinders must have a pressure of at least 1,800 pounds per square inch. Two or more cylinders, preferably at least four, should be used, if available.

(2) Charging capacity. Fully charged cylinders, if properly used in rotation, have capacity for charging pressure tanks approximately as follows:

    1 cylinder (used alone) . . . . . 2 pressure tanks
    2 cylinders (in combination). . 6 pressure tanks
    4 cylinders (in combination). . 24 pressure tanks
    5 cylinders (in combination). . 36 pressure tanks .
    6 cylinders (in combination). . 48 pressure tanks

(3) Apparatus. The apparatus for charging two pressure tanks by the use of cylinders consists of a filling line, two charging lines, and two cylinders. (Fig 23) The filling line and charging lines are obtained from the service kit. (Par 48) Plugs are provided to close off either half of the filling line when only one flame-thrower tank group is to be charged.

(4) Warning. Oxygen is sometimes shipped in cylinders having the same threads as nitrogen cylinders. If oxygen not mixed with nitrogen, as in air, is introduced into the fuel tanks of the portable flame thrower, a violent explosion may result. Therefore, the greatest care must be exercised to see that only air or nitrogen is used. Before a cylinder is connected, it should be tested to determine that it does not contain straight oxygen or some combustible gas. This may be done by introducing a burning splint into a jet of the contents. Oxygen causes the splint to burn quickly, whereas nitrogen extinguishes the flame. To make the test:

    (a) Fasten a thin splint of wood to a wire at least a foot long.
    (b) Ignite the splint.
    (c) Stand aside and hold it before the cylinder outlet.
    (d) Crack the valve slightly to permit a small stream of gas to emerge.
    (e) If the flame flares up, the gas is oxygen and MUST NOT be used.
    (f) If the gas itself catches fire, it may be hydrogen, acetylene, or some other combustible gas, which also must not be used.

(5) Attaching lines to cylinders. (Fig 23) The procedure for charging two flame thrower pressure tanks from two cylinders of nitrogen or compressed air begins as follows:

    (a) Remove the valve-protection caps from the cylinders.

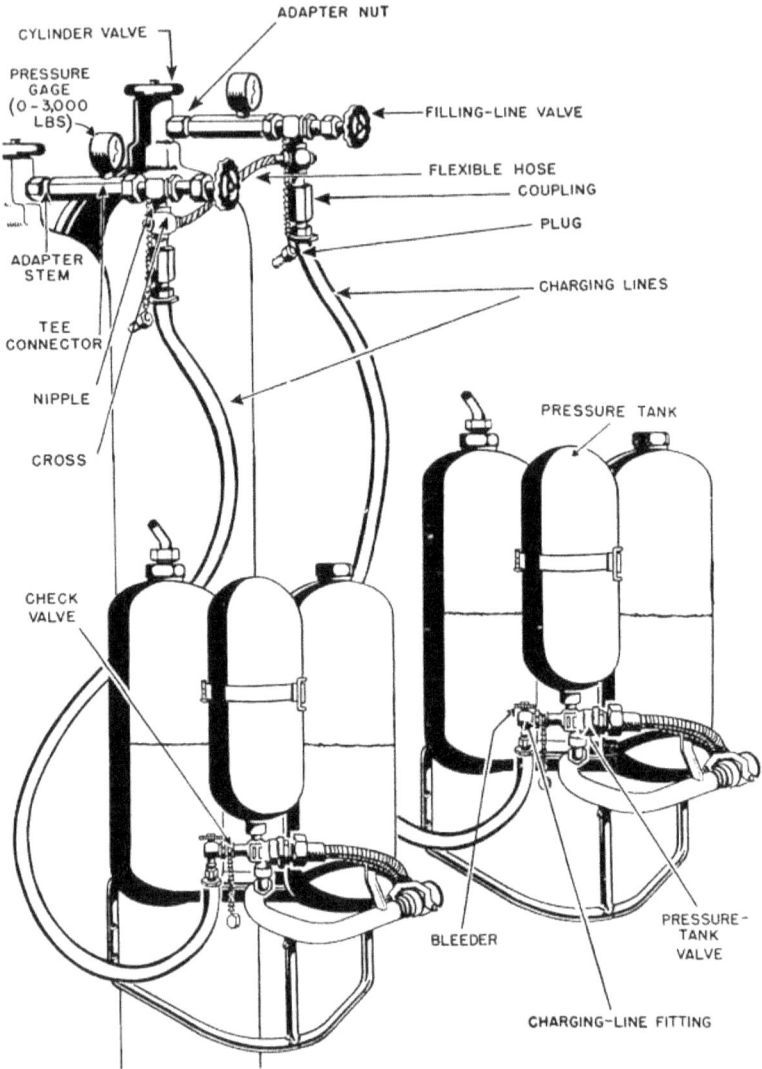

Fig 23. Charging two pressure tanks, using charging and filling lines, and cylinders of compressed air or nitrogen.

(b) Place the cylinders side by side with both outlets facing in the same direction. (If the ground is not level enough for the cylinders to stand up side by side, lay them horizontally with both outlets face up.)

(c) Before attaching the filling line to the cylinders, blow out dust. (Par 33) Then connect, using wrenches to make the joints pressure tight. Do not kink or bend the flexible hose. Cylinders must be close enough together to prevent strain on the flexible hose.

(d) Attach a charging line to each of the two couplings on the filling line.

(6) Attaching charging lines to pressure tanks.

(a) Close pressure-tank valves.

(b) Unscrew caps from check valves.

(c) Screw the charging-line fittings onto the check valves.

(d) Close bleeders.

(7) Charging. The operation of charging two pressure tanks from two cylinders is as follows:

(a) Close both filling-line valves.

(b) Open cylinder valves.

(c) Determine which cylinder has the lower pressure by the gages. Open the filling-line valve at the gage showing the lower pressure and fill the pressure tanks to the pressure shown by the gage. Close the valve. Then open the other filling-line valve and fill the pressure tanks until they reach pressures of at least 1,700 pounds per square inch as shown by the gage.

(d) When the pressure tanks have been filled, close the filling-line valves. Open the bleeders on the charging lines and leave them open until the pressure in the charging lines is released. Then close bleeders. Remove the charging-line fittings from the check valves. Screw the threaded caps on the check valves and tighten caps with a wrench.

(e) Repeat steps in (a) through (d) above for as many pairs of empty flame-thrower tanks as require charging.

(8) To insure proper pressure. Care should be taken to make certain that the compression delivered to the flame-thrower pressure tank is a full 1,700 pounds per square inch.

(a) If a filling-line valve leaks, tighten the packing nut on the valve with a wrench.

(b) When the higher pressure shown on the filling-line gages is less than 1,700 pounds per square inch, close the filling-line valve and the cylinder valve on the cylinder having the lower pressure. Remove and replace this cylinder with a fully charged cylinder. With chalk, mark the pressure on the cylinder which has been withdrawn.

(9) After charging. When charging has been completed:

(a) Close the filling-line valves. Observe the pressure indicated on each gage and mark the pressure on each cylinder using crayon, chalk, or pencil.

(b) Close the valves on the cylinders.

(c) Remove the charging-line fittings from the check valves, replace the threaded caps on the check valves, and tighten caps with a wrench.

(d) Remove the filling lines from the cylinders. Use two wrenches and take care not to twist or kink the flexible hose. Support the lines during the operation so that their full weight does not hang on the flexible hose during removal.

(10) Use of four-place lines. (Fig 24) The filling and charging lines found in two or more service kits may be combined for more

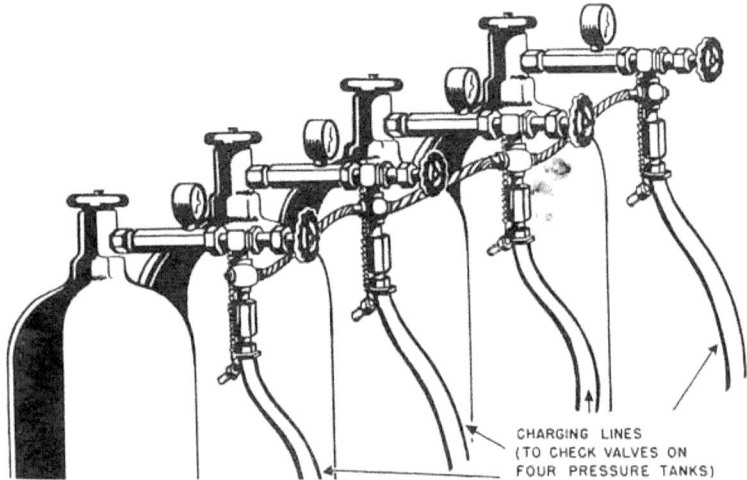

CHARGING LINES
(TO CHECK VALVES ON
FOUR PRESSURE TANKS)

Fig 24. Arrangement of cylinders and lines for charging four flame throwers. Flexible hose (assembly E81-3-6) from service kit is used to connect two filling lines.

efficient charging of large numbers of pressure tanks. An additional flexible hose is provided in each service kit for connecting two filling lines. The procedure for charging is similar to that described above for the two-place line. Air or nitrogen is taken first from the cylinder with the lowest pressure and last from the cylinder with the highest pressure. See a (2) above.

33. PRECAUTIONS WHEN PRESSURE-CHARGING.

Personnel will familiarize themselves with the following precautions:

a. Handling. Handle all cylinders and flame throwers carefully; never drop them and never subject them to shocks or blows. Keep valve-protection caps secured when cylinders are being handled, except when such handling is incident to the use of the nitrogen or air.

b. Storage. Keep all cylinders and charged flame throwers or tank groups (Par 77) in open or closed storage. They must, however, be protected from dampness and excessive rise in temperature caused by the direct rays of the sun or other source of heat. Avoid storing them near highly flammable substances, or in places where they may be struck by moving objects. Segregate empty cylinders to avoid confusion.

c. Personnel. Do not attempt to use compressed gases unless trained in this work. Use gases only for the purposes for which they are intended.

d. Cylinder valves. Do not tamper with safety devices in cylinder valves. If available, use the proper replacement parts for safety devices which are in need of repair. If such parts are not available, do not attempt to use makeshifts or nonstandard parts.

e. Opening of valves. Open valves slowly and fully each time nitrogen or compressed air is transferred from a cylinder. When a wrench is used, be sure it is one that fits properly, and that it is kept ready for instant use while the compressed gas is being released.

f. Threads. See that threads match before making connections. Some valves are provided with special threads which must be matched by the threads in the equipment being connected.

g. Correct equipment. Use gages, regulators, hose, pipe, and tubing of the type manufactured or specified for the particular apparatus or compressed gas.

h. Repair. Never attempt to alter or repair a cylinder.

i. Flames and sparks. Do not permit flames, sparks, or ignition from the flame thrower or other source to touch hose.

j. Blowing out dust. Immediately before coupling an attachment to the pressure tank or cylinder valve, open it for an instant to blow out any dust or dirt. Never stand where gas or dirt may be blown into the eyes or face. If the valve is difficult to open, apply more force gradually.

k. Special devices. Do not attempt to use any special connections or equipment without the approval of a qualified expert.

l. Keeping valves closed. Keep the valve of each cylinder closed when its contents are not actually being released from or admitted to the cylinder. This applies alike to all cylinders, whether they contain a compressed gas or are empty.

34. CHARACTERISTICS OF FUELS.

Thickened fuels give up to twice the range of liquid fuels. The stream of thickened fuel is comparatively narrow. Most of the glue-like fuel clings to and burns in or on the target for as long as 6 minutes. Liquid fuels, on the other hand, are largely consumed in flight to the target. If the location of small openings in the target is known, the stream of thickened fuel can be spotted by accurate aiming so that most of the fuel enters directly into the openings. While it does not billow around corners as does liquid fuel, thickened fuel strikes the target with force enough to ricochet inside. It clings to skin and clothing while burning. It also has excellent incendiary effects. The initial flame and smoke are less from thickened fuel than from liquid fuel, but the lower visibility, greater range, and much longer burning period of thickened fuel compensate for its smaller screening effect. Liquid fuels are easier to pour when filling than are thickened fuels.

35. PREPARATION OF THICKENED FUELS.

a. Ingredients. Thickened fuels consist of U.S. Army fuel thickener mixed with fuel.

(1) Thickener. U.S. Army thickener is supplied in airtight cans, each containing 5-1/4 pounds of the material.

(2) Gasoline and fuel oil. Gasoline alone is often used with thickener, but mixtures of gasoline and light fuel oil may be used satisfactorily. The light fuel oil can be either No. 1 fuel oil, No. 2 fuel oil, automotive diesel oil, or kerosene. These mixtures give more heat and do not form crusts. Except in hot climates, 75 percent or more of the mixture by weight or volume should be gasoline. (If too much light fuel oil is included, the fuel tends to separate into two layers.) In tropical theaters, a thickened blend of 50 percent gasoline and 50 percent light fuel oil has been reported to give favorable results. Storage qualities are not known, however. Another mixture which has been well recommended in field reports is 15 gallons of gasoline to 5 gallons of diesel fuel oil. Issue gasoline may be used, but locally procured gasolines which contain alcohol are not suitable.

b. Proportion of thickener to fuel. Less thickener is recommended than formerly. A low ratio of thickener gives a thickened fuel with many of the characteristics of liquid fuel. One can of thickener to 20 U.S. gallons of gasoline, or gasoline and light fuel-oil mixture, gives good results. This is a 4.2 percent by weight mixture. Except in hot weather, a fuel mixture of less than 3 percent thickener requires such long stirring that its preparation is impractical.

c. Equipment. An open-head 55-gallon or 42-gallon drum and an improvised wooden mixing paddle are used. Five-gallon cans may be employed to transfer the ingredients. The paddle should be approximately 5 feet long, 2 inches wide, and 1 inch thick. If a standard 55-gallon, open-head drum with an internal diameter of 27-7/16 inches is used, the improvised paddle should be marked to indicate gallons as follows:

Gallons	Inches
40.	23-1/2
20.	11-3/4

Do not use a metal paddle because of the danger of striking a spark from the drum. Never use galvanized containers for mixing and storing thickened fuels. These may cause the fuel to break down and become excessively thin. An improvised funnel may be helpful in filling drums with prepared fuel for aging or transporting.

d. Temperatures. (1) Below 50 degrees. If the temperature is below 50 degrees Fahrenheit, it is helpful to prepare thickened fuel indoors, in a heated room. All precautions should be particularly observed. (Par 40)

(2) Above 90 degrees. When the fuel is hotter than 90 degrees Fahrenheit, the thickener reacts very rapidly. In this case, it is easier to prepare batches of 20 gallons each, but any number of batches may be prepared in succession.

e. Moisture. (1) Effect of moisture. Water in thickened fuel breaks down or reduces the viscosity of the gel and thereby reduces the range of the flame thrower. This effect may not be noticeable at once, but the stability of the fuel is affected.

(2) Dryness of thickener. Dry thickener is extremely hygroscopic, that is, it absorbs moisture from the atmosphere very rapidly. For this reason, thickener is shipped in hermetically sealed tin cans containing the exact quantity of powder required for mixing with 20 gallons of fuel to prepare a 4.2 percent mixture. It is important that the gasoline or fuel oil and gasoline be measured out before the thickener container is opened. The powder then should be poured immediately into the liquid.

(3) Dryness of containers. It is important that all containers used in mixing and handling the fuel be dry.

(4) Keeping water out of gasoline. Gasoline, especially when it has been stored in vented containers, frequently includes free water. Therefore, when using gasoline from a bulk-storage tank or an open drum, first place it in a clean, dry drum; allow it to stand quietly for at least an hour; then carefully pour off the gasoline from the top and discard the last gallon or two.

f. Pouring and stirring. (Fig 25) The liquid fuel is poured into the open drum; a pail or a paddle (Par 35 c) is used for measuring. One man then stirs the fuel vigorously. Another takes a can of thickener, splits it with a machete, bayonet, or ax, and pours it immediately into the fuel. Any large lumps of powder are broken by hand before the powder is added to the fuel. When mixing 40 gallons at a time, the two cans of thickener should be opened and added to the fuel in rapid succession. If the contents of the first can are permitted to gel before adding the second can, it will be difficult to obtain a uniform mix. Continue to stir.

g. Examining fuel. Lift the paddle quickly. If the mixture drops or runs from the paddle, additional stirring is necessary. When the paddle comes out clean, except for an adhering film, stirring should be stopped, provided there is no further visible

Fig 25. Measuring fuel ingredient into mixing drum. Paddle for measuring and stirring is improvised.

Fig 26. Transferring newly mixed thickened fuel from mixing drum to storage or shipping container for aging.

settling of particles of thickener.

h. Loading shipping drums. When stirring is completed, the mix is immediately bucketed (Fig 26) through a funnel into the shipping drum. The second bung hole should be open, if possible, to provide a vent to aid in pouring. Two men do the bucketing, each handling one pail so that the funnel may be kept loaded with mix and the shipping drum filled as rapidly as possible. Finally, the open-end drum should be picked up and its contents poured into the funnel. Not more than 50 gallons of thickened fuel should be loaded into a 55-gallon drum. The funnel should then be removed and replaced by a plug. The vent opening of the drum should also be closed. (See Paragraph 39 for pressure method of filling storage drums.)

i. Unused thickener. Any thickener remaining in opened cans should be discarded. Since moisture in the air can quickly ruin its properties, no attempt should be made to save it.

j. Aging and storing. Newly mixed fuel has the appearance of tapioca pudding. (Fig 27) It should preferably be stored overnight before use. It may, however, be fired within 1 hour after mixing. To keep fuel in good condition, drums for shipping and storing must be clean, moistureproof, dry, strong, and unrusted, but not galvanized. They must be kept tightly closed and should be laid on their sides so that rain water will not collect around the bungs.

k. Testing fuel. Before use on missions, all fuels should be tested by being fired from a flame thrower. This is advisable because the characteristics of the fuel ingredients often vary.

Fig 27. Contrasting newly mixed thickened fuel (right)
with aged fuel (left).

36. PREPARATION OF LIQUID FUELS.

a. Choice of ingredients. Thin fuels are easy to ignite, but they lack range and are largely burned in flight before reaching the target. For this reason, liquid fuels should contain the lowest proportion of gasoline and the highest proportion of heavier oils that permits easy ignition. In hot climates, less gasoline is needed than in cold climates. Exactness of proportion, however, is not of great importance. Suitable blends are as follows:

(1) Equal parts by weight or by volume of gasoline, light fuel oil, and heavy (bunker) fuel oil. The light fuel oil can be either No. 1 fuel oil, No. 2 fuel oil, automotive diesel oil, or kerosene.

(2) One part gasoline to four parts of cleaned crankcase drainings. (Par 36e) Unused motor lubricating oil can be employed in place of crankcase drainings, but usually it will be unavailable for flame-thrower use.

b. Preparation of ingredients. Before mixing blends, the following steps should be taken:

(1) Gasoline, diesel oils, and fuel oils. These fuel materials should be allowed to stand quietly for at least 30 minutes to permit any small quantity of water present to settle to the bottom. When transferring the fuel to another container, remove the fuel carefully so that no water is remixed with it.

(2) Crankcase drainings. If possible, crankcase drainings should be allowed to stand quietly in a container for at least 1 day. When pouring, take care to prevent the transfer of any of the sludge which may have settled in the bottom of the container.

c. Equipment. An open-head 55-gallon or 42-gallon drum and an improvised wooden mixing paddle are used. The paddle should be approximately 5 feet long, 2 inches wide, and 1 inch thick. A metal paddle should not be provided because of the danger of striking a spark from the drum. Five-gallon cans may also be furnished for measuring and transferring ingredients. Clean, unrusted, steel storage drums should be at hand. They should be at least 16-gage to have sufficient strength to withstand the internal vapor pressure of the fuel.

d. Stirring. All the ingredients should be stirred in the drum with the paddle until they appear to form a uniform mixture. This should require approximately 2 minutes.

e. Crankcase-draining blends. If crankcase drainings are used as an ingredient (Par 36b), it is preferable to allow the prepared mixture to settle for 24 hours after stirring, because the gasoline in the mixture may cause additional sludge to be deposited. Even after this settling period, it is recommended that the mixture be poured through cheesecloth or some similar fabric before the flame thrower is filled. Crankcase-draining blends should be allowed to stay in the flame thrower only long enough for completion of a mission, because additional sludge which may form from standing will clog the weapon.

f. Transferring. The mixture should be transferred either directly into the flame-thrower fuel tanks (Pars 37 through 40) or into storage drums. (Par 35h)

g. Emergency mixing in fuel tanks. In an emergency, mixing can be done in the flame-thrower fuel tanks by adding the ingredients in correct proportions and then shaking or stirring.

h. Testing fuel. Before fuel is used on a mission, it should be tested, if possible, by being fired from a flame thrower.

i. Storage. Fuels may be used immediately after preparation. If the blend contains crankcase oils, the fuel should be fired as soon as practicable after filling. Other liquid blends may be stored indefinitely until required for use. For storage precautions see Paragraph 40. The storage drums also should be kept tightly closed to prevent loss of gasoline through evaporation and to prevent moisture from entering the fuel. If stored in the open, the drums should be laid on their sides so that rain water will not collect adjacent to the bungs. An unrusted and undamaged 16-gage or 18-gage drum has sufficient strength to withstand the internal vapor pressure of the fuel.

37. FILLING BY POURING.

(Fig 28) This method is the simplest and quickest for liquid fuel,

Fig 28. Filling fuel tanks by pouring. Any clean container may be used. A funnel may be improvised.

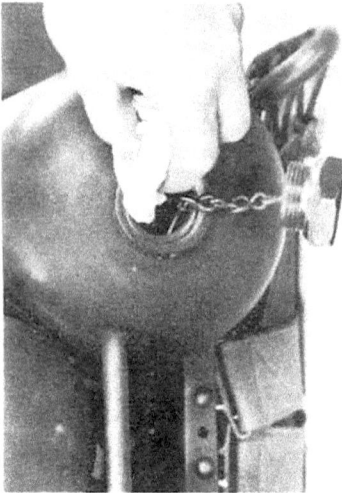
Fig 29. Wiping plug seat.

but it may be too slow for some thickened fuels. The procedure is as follows:

a. Stand the tank group on the ground or a platform. If the tank group is not connected to the gun group, lock coupling plug in tank coupling. (Par 70)

b. Using a 1-3/4-inch wrench, unscrew the filling plug and the safety-head plug.

c. Inspect interior of tanks to see if clean and free from foreign matter. If not clean, flush with gasoline.

d. Using an improvised funnel, fill to within 2 inches of the top of both plug openings. This allows sufficient void. The tanks will then contain approximately 4 gallons of fuel.

e. Wipe the fuel-tank plug seats and the plug threads with a clean, dry cloth. (Fig 29) If plug has a tendency to freeze to seat, lubricate (Par 49 b) before screwing in the filling and safety-head plug assemblies. Tighten with wrench.

f. Wipe any spilled fuel from weapon.

38. FILLING BY FORCE PUMP.

A force pump, if available, may be installed with a short length of pipe in the top opening of a drum of fuel for filling flame-thrower fuel tanks. Keep working parts of pump clean.

39. FILLING BY BLOWING.

Thickened fuel may be readily forced into the fuel tanks of flame throwers by the use of extremely low pressures of compressed air or nitrogen. Flame thrower fuel filling kit E6 or equivalent may be used. When equipment is available, filling by blowing is more efficient for filling large numbers of flame throwers with thickened fuel. Pouring or pumping are more time-consuming, depending on the consistency of the gel. The consistency may vary among batches even when the same proportion of thickener is used. The amount of moisture in the fuel seems to cause this variation. The precautions listed in Paragraph 40 should be observed.

a. Source of pressure. When the pressure in cylinders of compressed air or nitrogen has fallen too low to be of further use in filling pressure tanks of flame throwers, the remaining pres-

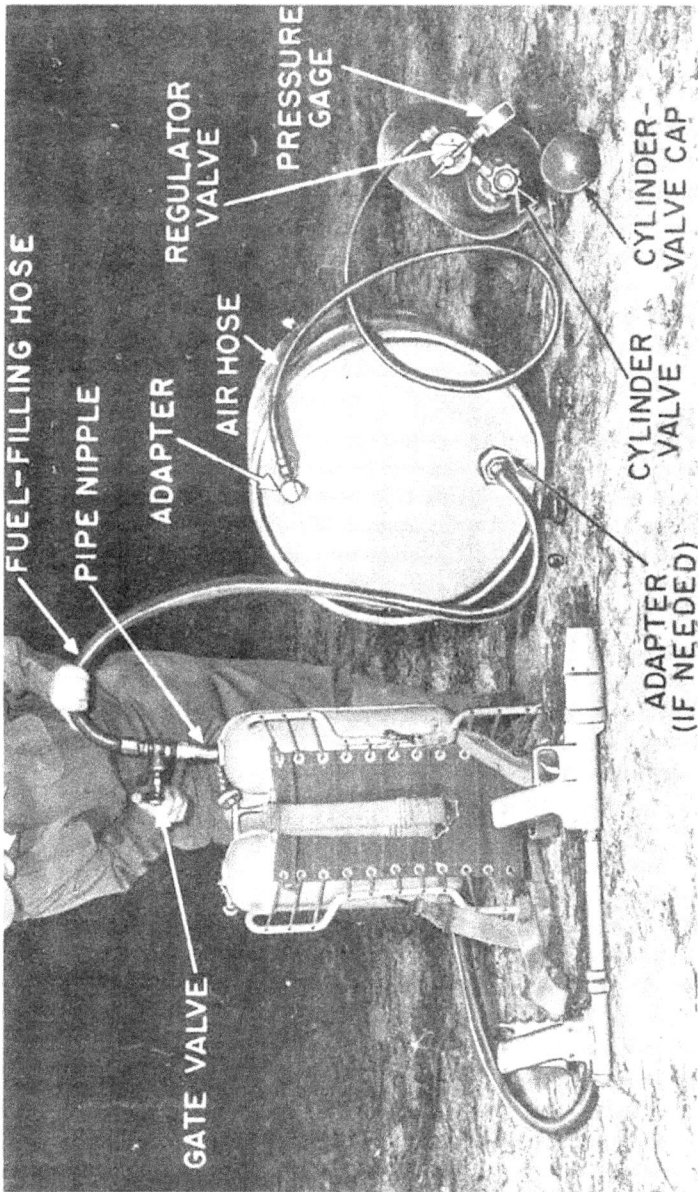

Fig 30. Blowing thickened fuel into fuel tanks by use of cylinders of compressed air or nitrogen.

sure may be used to blow fuel into fuel tanks if the regulator valve can reduce pressure down to 20 pounds per square inch. For precautions, see Paragraph 33. An air compressor or a hand air pump (tire pump) may be used in place of a cylinder if the latter is not available. Pressure of no more than 15 to 20 pounds per square inch should be used on the fuel drums. Only a diaphragm-type regulator valve can be used safely. This valve must be capable of regulating any pressure that may be applied to it.

b. Drums. Clean, noncorroded, steel, 55-gallon drums should be used. Drums of United States manufacture which meet requirements will be stamped ICC-5 or ICC-5A, followed by three numbers in sequence, for example, "14-55-44." The number "14" indicates the gage of the metal; "55" indicates the capacity in gallons; and "44" indicates the year of manufacture. A steel drum of 14 gage, or heavier, is preferable, but lighter drums (of 16 or 18 gage) may be used. Drums made of gages lighter than 18 gage (20- or 22-gage) must not be used. Drums should never be moved while under pressure.

c. Connections. The source of pressure (see a above), the drum of fuel, the fuel-filling line, the air hose, and other parts, are connected as shown in Figure 30. Threaded adapters are used, as necessary, to fit lines to the drum. All threaded connections should be made tight by the use of wrenches on the joints. The drum and the pressure cylinder (if the latter is used) should be laid on their sides on the ground or a platform. The opening of the drum connected to the fuel-filling hose should be close to the ground or platform. If tank group is filled without gun group, lock coupling plug (Par 70) in tank coupling.

d. Procedure. To fill fuel tanks:

(1) Remove both the filling and safety-head plugs.

(2) Inspect interior of tanks to see if clean and free from foreign matter. If not clean, flush with gasoline.

(3) Place end of fuel-filling hose in either one of the two fuel-tank plug holes, using a nipple as a spout.

(4) Start air compressor or pump, or open the valve on the cylinder of compressed air or nitrogen. Open regulator valve on filling line by turning handle slowly until gage shows 15 to 20 pounds pressure, but no more. Caution: "Cracking," or opening a cylinder valve without using the proper regulator valve (Par 39 a), may result in explosive pressure in the drums.

(5) Both tanks must be filled to within 2 inches of their tops. Close valve on fuel-filling hose to halt flow at this level.

(6) If no additional flame throwers are to be filled, close pressure-cylinder valve, or stop compressor or pump. Then, using wrench, slightly loosen the air line at the drum, allowing pressure to bleed. When the pressure in the drum has fallen to that of the atmosphere, close regulator valve.

(7) Roll drum slightly and gently until fuel-filling hose is at top of drum.

(8) If there are valves on each end of the fuel-filling hose, use wrench to slightly loosen hose, allowing gradual escape of pressure. Stand away from, and at the side of, the connection. Keep hose pointed away from other personnel. When all pressure has been released, complete unscrewing of hose.

(9) Wipe fuel-tank plug seats and the plug threads with a clean, dry cloth. Then screw in filling plug and safety-head plug assemblies, applying grease (Par 49 b) if plug tends to freeze to seat. Tighten with wrench. Wipe any spilled fuel from weapon.

40. PRECAUTIONS WITH FUELS.

a. Flammability. All fuels used in flame throwers obviously are highly flammable and must be handled, stored, and used with extreme care. Diesel oil, fuel oil, and kerosene require the same care as does gasoline.

b. Indoor storage. When it becomes necessary to handle gasoline in a room or building, the windows and doors should be open and care taken that no unprotected flame which might ignite the fumes is in the vicinity. The doors and windows should remain open for a sufficient length of time afterward to allow any vaporized gasoline to escape.

c. Flames and sparks. The presence of open flames, heated stoves, electrical tools and apparatus, and other equipment likely to cause sparks must not be permitted. Even nails and metal cleats in shoes are a potential hazard in the presence of combustible fumes.

d. Smoking. "No Smoking" signs must be posted in prominent places about the premises and the rule against smoking must be strictly enforced.

e. Ventilation and cleaning. The buildings in which fuel is stored or used must be well ventilated and thoroughly cleaned every day. No rubbish or other flammable material should be permitted to remain in or near such buildings.

f. Spillage. Care should be taken that fuel is not spilled. Any spillage should be removed promptly.

g. Safety cans. Safety cans should be used, if possible, for storing small quantities of gasoline, as they have covers that must be forcibly held open to remove or add gasoline.

h. Rags. Metal receptacles with metal lids should be provided for discarded, oily, or gasoline-soaked rags. These rags must be disposed of daily.

i. Electrical apparatus. Vaporproof incandescent electric lamps, switches, and other appliances of approved type should be used. Open switches, relays, and similar apparatus, or motors with commutators, must not be used where gasoline fumes may be encountered.

j. <u>Hose</u>. Flexible metal, rubber, and rubber-metal hose should be inspected regularly (at least four times a year) and discarded when noticeably deteriorated.

k. <u>Toxic fumes</u>. Gasoline fumes are somewhat toxic and should not be inhaled.

l. <u>Leaks</u>. Leaks must never be neglected, and the fact that gasoline is a dangerous liquid must always be kept in mind. Inspections for leaks should be made frequently, particularly at pipe and hose joints.

m. <u>Fire extinguishers</u>. Carbon tetrachloride, carbon dioxide, or foam-type fire extinguishers should be provided and located where they will be accessible in the event of fire. Sand, not water, should be thrown on burning fuel if suitable extinguishers are not available.

n. <u>Leaded gasoline</u>. Gasoline often contains a poisonous lead compound. Such gasoline, or fuel containing leaded gasoline, should not be allowed to touch the body, especially the lips, eyes, open cuts, and sores.

Section IX OPERATION
UNDER UNUSUAL CONDITIONS

41. WET CONDITIONS.

The M2-2 flame thrower may be carried and fired successfully in the rain or even after short immersion in water. After use when wet, it should be dried to prevent rusting, cleaned, and lubricated. (Pars 49 and 55) Areas where paint has worn off should be touched up with fresh paint. The weapon should be stored in a dry place. Moisture must not be allowed to enter fuel, ingredients of fuel, or containers of ignition cylinders.

42. DUST AND MUD.

Keep all possible dust, earth, and mud out of the flame thrower; particles may interfere with the operation of spring case, valves, bearings, and pressure regulator. Store weapons and auxiliary equipment in closed chests and boxes when not in use. (Par 77) Clean before use. (Pars 51 and 52)

43. HEAT.

A hot climate or exposure to the sun makes the fuel thinner when in containers. Thin fuel has shorter range; it is largely consumed in the air before it reaches usual effective ranges. Where the climate is torrid, less gasoline or other thinning agents should be used in a fuel blend than normally. (Pars 34 through 36)

44. COLD.

Cold weather reduces total heat produced at target but seldom enough to seriously lower value of a firing mission. Incendiary effects may be decreased because materiel is less flammable when cold. The weapon may be used at temperatures as low as minus 20 degrees Fahrenheit. To improve ignition, use more gasoline in fuel than normally. (Pars 34 through 36)

45. WIND.

Flame throwers should not be fired into strong head winds or across strong side winds. (Par 23)

Section X DEMOLITION
TO PREVENT ENEMY USE

46. DESTRUCTION PROCEDURE.

If circumstances should force abandonment of chemical warfare materiel in the field, it is destroyed or rendered useless to prevent its use or study by the enemy. The following methods are recommended:

a. Flame thrower. One or more small-arms bullets through the fuel tanks will prevent any immediate use of the flame thrower. Additional rounds may be put through the pressure tank. If the pressure tank is charged, the pressure-tank valve should be opened for a few seconds, thus permitting the contents to dissipate. This is necessary if rounds are to be fired point-blank. The gun may be rendered useless by bending it over a hard object. A sledge or ax will demolish valves and tubes. A fragmentation grenade will also achieve demolition.

b. Filling and charging apparatus. The flexible tubing, gages, and valves may be destroyed by blows with an ax, sledge, or other heavy instrument. The large pressure cylinders are rendered useless by releasing the contents and then destroying the valves by blows with an ax or sledge. Cylinders can be stacked like cordwood in groups of five and demolished by the detonation of four 1/2-pound blocks (2 pounds) of TNT in their midst. The air compressor may be destroyed by a similar procedure.

c. Fuel. Burn.

d. Mixing apparatus. Containers and filling lines may be rendered useless by ax or sledge blows, or by small-arms fire.

e. Thickener. Cans of thickener should be broken open. Contents should be thrown into a fire or into a body of water.

f. Ignition cylinders. Burn to destroy. Personnel should stay several yards from the fire because the cylinders ignite with a slight detonation.

PART THREE
MAINTENANCE INSTRUCTIONS

Section XI   GENERAL

47. SCOPE.

Part Three contains information for the guidance of the personnel of the using organizations responsible for the maintenance
(1st and 2nd echelon) of this equipment. It contains information
needed for the performance of the scheduled lubrication and
preventive maintenance services as well as descriptions of the
major systems and units and their functions in relation to other
components of the equipment.

Section XII SPECIAL ORGANIZATIONAL
TOOLS AND EQUIPMENT

48. SERVICE KIT.

One service kit for portable flame thrower M2-2 will be
furnished for each six M2-2 portable flame throwers. The kit
includes tools, equipment, and spare parts for second echelon
maintenance and for pressure-tank charging. Adjustable wrenches
may be included in place of the plain-end wrenches listed. Numbers
listed with items are Chemical Warfare Service stock numbers.
Approximate contents are as follows:

   a. Tools.

      1 Screwdriver, cabinet, 4-1/2-inch blade length, 3/16-inch
         blade diameter, H22-50-13. (Fig 8)

      1 Screw driver, common, 6-inch blade length, 5/16-inch
         blade diameter, H22-50-6. (Fig 8)

      2 Wrenches, hex, 3/16 inch across flats (for 3/8-inch
         socket-head set screws), H22-49-91.

      2 Wrenches, hex, 1/8 inch across flats (for 1/4-inch
         socket-head set screws), H22-49-12. (Fig 8)

1 Wrench, valve-adjusting, assembly A81-6-48. (Fig 8)
1 Wrench, heavy "S", 1-3/8-inch and 1-1/2-inch openings, 12 inches approx length, H22-49-113. (Fig 8)
1 Wrench, engineers', double head, 3/4-inch and 7/8-inch
openings, 9 inches approx length, H22-49-115. (Fig 8)
1 Wrench, heavy "S", 1-3/8-inch and 1-3/4-inch openings, 12 inches approx length, A81-6-49. (Fig 8)
1 Wrench, engineers', single head, 1-1/8-inch opening,
10-1/2 inches approx length, H22-49-31. (Fig 8)
1 Wrench, adjustable, single end, 6 inches approx length
(crescent-type), H22-49-67. (Fig 8)
b. Accessories and spare parts.
1 Line, filling, pressure cylinder, assembly C81-3-4. (Fig
23)
1 Hose, flexible, assembly E81-3-6. (Fig 24)
2 Lines, charging, pressure cylinder, assembly B81-3-29.
(Fig 23)
1 Tank and valve, pressure, assembly (less shaft and handle)
B81-1-374. (Fig 33)
1 Shaft, flexible, valve, assembly E81-1-470. (Fig 33)
1 Handle, valve, A81-1-473. (Fig 33)
1 Nut, machine-screw, hex, 5/16-inch, 24NF-2, H22-93-55.
(Fig 33)
2 Case, spring, assemblies B81-1-444. (Fig 9)
2 Diaphragm, valve, assemblies A81-1-416. (Fig 9)
1 Hose, fuel, flame thrower, M1, assembly B81-1-498.
(Fig 48)
2 Plugs, coupling, E81-1-514. (Fig 7)
6 Heads, safety, R81-1-561. (Fig 39)
1 Gage, fuel tank testing, assembly E81-6-57. (This assembly includes a plug drilled, tapped, and fitted with a
0-500-pound pressure gage.)
3 Washers, coupling, A81-1-513. (Fig 9)
2 Cord, cotton, seine, No. 4 hard braided, mildewproof,
O.D., (1/8-inch diameter by 25-feet skeins), H100-4-5.
6 Bushings, pipe, head, 3/4 inch by 1/2 inch, (galvanized
iron), H98-5-93. (Fig 9)
1 Regulator, pressure, assembly B81-1-438. (Figs 33 and
37)
1 Compound, anti-seize, white lead base, (for threaded fittings) 1/4-pound can, H99-3-12.
2 Gages, pressure cylinder testing, assembly B81-6-90.
(Fig 32)
1 Catalog CW7-440114, Army Service Forces, "Portable
Flame Thrower M2-2."
1 Technical Manual 3-376A, "Portable Flame Thrower
M2-2."

WAR DEPARTMENT LUBRICATION ORDER NO. 4001

WAR DEPARTMENT, Washington 25, D. C. (5 May 1944)

PORTABLE FLAME THROWER M2-2

For detailed instructions, refer to TM 3-376A

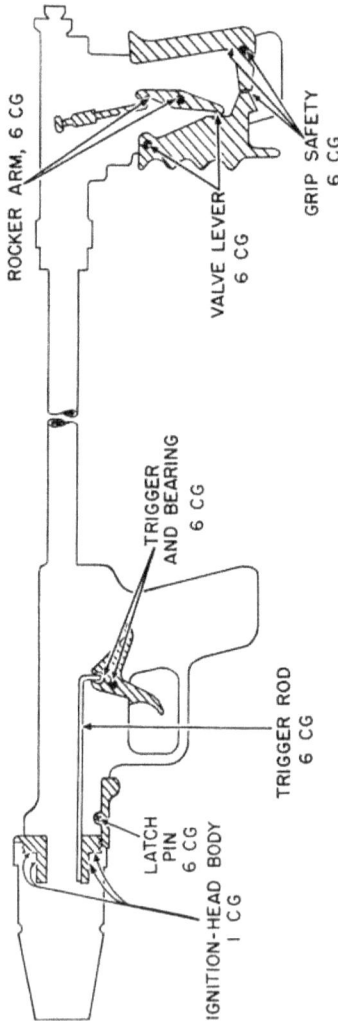

ROCKER ARM, 6 CG

GRIP SAFETY
6 CG

VALVE LEVER
6 CG

TRIGGER
AND BEARING
6 CG

TRIGGER ROD
6 CG

LATCH
PIN
6 CG

IGNITION-HEAD BODY
1 CG

——— KEY ———

LUBRICANTS	INTERVALS
CG – GREASE, GENERAL, PURPOSE NO.1 (ABOVE + 32°F) NO. 0 (BELOW + 32°F)	1– AFTER EACH MISSION 6– AFTER SIX MISSIONS OR MORE OFTEN

THIS ORDER IS TO BE FASTENED TO INSIDE LID OF FLAME THROWER PACKING CHEST.

To requisition a replacement Lubrication Order address Office of the Chief, Chemical Warfare Service, Washington 25, D. C.

NOT TO BE REPRODUCED in whole or in part without permission of the Office of the Chief, Chemical Warfare Service.

NO. 4001

Copy of this Lubrication Order will remain with the equipment at all times; instructions contained therein are mandatory and supersede all conflicting lubrication instructions dated prior to 5 MAY 1944

By order of the Secretary of War:
G. C. Marshall, Chief of Staff.

Official:
J. A. Ulio,
Major General,
The Adjutant General.

Fig 31. Lubrication order.

Section XIII LUBRICATION

49. LUBRICATION.

a. Gun group. War Department Lubrication Order No. 4001 (Fig 31) shows the parts which require lubrication, the lubricants, and the intervals.

(1) Lubricants. Grease, general purpose, No. 1 is used, except for temperatures below freezing, when grease, general purpose, No. 0 is used. The bearing surfaces should be lightly coated with the grease.

(2) Frequency of lubrication. The surfaces of the ignition-head body which touch the spring case should be lubricated after each use of the weapon. Other parts are lubricated after six firing missions, six training sessions, or oftener. All should be thoroughly cleaned (Pars 52, 55, and 56) with gasoline, dry cleaning solvent, or other solvent, then dried before lubrication. If the gun is disassembled for any other reason, it should be lubricated before reassembly.

(3) Records. To ascertain when six missions have been fired, a record of firing (Par 2) should be kept with each flame thrower.

b. Tank group. The tank group ordinarily requires no lubrication. However, it may under the following exceptional circumstances:

(1) If the tank group has been immersed in water for several hours, the flexible shaft of the pressure valve may have lost its lubricant. If so, remove the shaft (Par 66 b) and inspect. If lubricant is not present, as indicated by difficulty of movement after removal, dip the shaft in solvent to clean and then dip in warmed grease, general purpose, No. 1. Replace shaft in valve.

(2) If filling or safety-head plugs (Figs 39 and 40) tend to stick to fuel tanks, apply grease, general purpose, No. 1 (No. 0 if below freezing temperatures) before replacing plugs.

Section XIV PREVENTIVE
MAINTENANCE SERVICES

50. GENERAL.

Preventive maintenance services, as prescribed by Army Regulations, are a function of using organization echelons of maintenance. These services consist of:

a. Before, during, and after operation services performed by the firers and assistants.

b. Scheduled services performed by organizational maintenance personnel (service when filling and charging, and service after six firing missions).

51. BEFORE-OPERATION SERVICE OF TANK GROUP.

The following services are to be performed before filling, charging, and loading the flame thrower with pressure, fuel, and ignition cylinder:

a. Pressure-tank valve. Open and close pressure-tank valve to test for ease of operation.

b. Threaded connections. Check all threaded connections for tightness, using appropriate wrenches.

c. Tank coupling. Examine coupling for cleanliness and ease of movement of lock and cams. (Par 70) Clean if necessary. If washer is broken, replace, using screw driver to pry out.

d. Plugs. Check filling plug and safety-head plug for completeness of parts (Par 69 a) and cleanliness of threads and seats. Clean, if necessary, with cloth. If rod or rod and chain have broken off and fallen in tank, turn tank upside down and remove. Remove deflector tube from head (using hand, not wrench). Inspect to see if diaphragm is intact. If diaphragm is ruptured, replace the safety head with an unbroken head. (Par 69 b, c) Reassemble plug, head, and deflector tube in left fuel tank. (Fig 11) Tube should face to rear and at a 45-degree angle to operator's left shoulder. (Fig 18) Screw in deflector tube by hand; do not use wrench on deflector tube. Tighten lock nut with wrench.

e. Pressure-tank clamp. The clamp should hold the pressure tank tightly in place. If tank is loose, a wooden splint or wedge under the clamp may be used as a temporary expedient.

f. Carrier-frame bolts. Check tightness. Use wrench.

g. Carrier. (Par 71) Examine all canvas, webbing, and cord for signs of mildew, rot, or wear. Replace defective parts. Move flame thrower to dryer storage if mildew occurs.

h. Cord (lashing). Check for tightness. If necessary, make tighter and use secure, slip-proof knots. When the tank group is filled with fuel and adjusted on the firer, its weight should be carried chiefly by the canvas and webbing, not by the metal frame.

i. Shoulder and body straps. Adjust straps to fit firer. (Pars 19 and 71) A loose tank group can cause discomfort or injury when the wearer changes positions while on a mission. Check presence and condition of the two pins and two cotter pins which hold shoulder straps to steel support. Check fasteners.

52. BEFORE-OPERATION SERVICE OF GUN GROUP.

The following services are to be performed before filling, charging, and loading with pressure, fuel, and ignition cylinder:

a. Hose nipple, tank end. Examine to be sure nipple is clean and not badly nicked. If badly nicked, the nipple may not make a tight seal at the tank coupling. A leak and loss of pressure may result. See Paragraph 73 d for repair of the nipple.

b. Fuel hose. Examine surface of hose for cracks or other signs of deterioration. Special attention should be paid to portions adjacent to the gun and tank coupling, which are subjected to severe flexing. If hose is defective, replace. (Par 73 b, c) Do not patch.

c. Hose nipple, gun end. Check tightness of threaded connection between hose and fuel-valve body, using hand or very light wrench pressure.

d. Shield. Remove ignition shield. Check cleanliness of threads on shield and on ignition-head body. If not clean, use cloth. When reassembling (Par 18 c), shield should turn freely until it locks in correct position.

e. Valve lever and needle. (1) There should be some play in the valve lever. To test, remove ignition shield. Compress grip safety and valve lever slowly, observing the motion of the valve needle. The valve lever should move approximately 1/16 inch before the needle begins to move.

(2) Valve needle should be seated firmly in the barrel nozzle. After the valve lever is pulled back and released, no play should occur in the needle. For adjustment of needle, see Paragraph 75 d.

f. Screws. Use screw driver to test tightness of all screws.

g. Spring retainer and plug. Check tightness of spring retainer and plug (Fig 47) by using hand or very light wrench pressure.

h. Ignition head. All exposed surfaces of the shield, nozzle, needle, and other parts of the ignition head, or adjacent to it, should be clean. If not, use cloth.

i. Atomizer hole. With the fuel valve held fully open, insert a fine wire in the atomizer hole of the nozzle to clean the hole. Then use cloth-wrapped splint to remove from the inside of the nozzle any foreign matter pushed through the atomizer hole. If such matter is not removed, it may interfere with the closing of the fuel valve needle at the nozzle. Repeat procedure in e (2) above.

j. Spring case. Spring case should turn freely on ignition head. If it does not, clean any grease or dirt from surfaces with cloth and relubricate. (Par 49)

k. Trigger. Pull trigger once or twice to find whether it operates easily and whether it returns to position. If not, clean and lubricate trigger. (Par 49) Check condition of trigger spring.

l. Trigger rod. Check position of the trigger rod when trigger is pulled back all the way as when firing. The rod should extend approximately 1/16 inch beyond the end of the lug in the ignition head. If it does not, bend the rod slightly, reverse position of bearing, or replace worn parts.

53. SERVICE WHEN FILLING AND CHARGING.

a. Inspection of fuel tanks. Just before filling and charging, remove plugs (Par 69 b) and examine interior of fuel tanks to see whether they are clean and free from foreign matter. If not clean, flush with gasoline until clean.

b. Fuel level. When filling (Pars 37 through 40), see that fuel reaches the same level in both tanks. If leveling does not occur, the tank connector may be clogged with foreign matter. If so, clean, as in a, above. After filling, wipe plug seats with a cloth before replacing plugs. Wipe any spilled fuel from weapon.

c. Pressure-tank valve. Before charging the tank group with air or nitrogen, open and close the pressure-tank valve several times by hand to be sure it operates freely. If it does not, adjust as described in Paragraph 66 d.

d. Testing for leaks in pressure system. After charging, and as few hours as possible before a mission, use an 0-3,000-pound gage furnished in service kit to test pressure. (Fig 32) To install gage, unscrew check-valve cap and screw gage in check-valve body. If pressure has fallen below that to which the tank was charged (Par 32), a leak is indicated. Remove gage, replace check-valve cap, and check for leaks at joints between pressure tank and valve and between tank valve and check valve. (A wrench should be used to tighten cap on check-valve body so as to avoid producing an additional leak.) Large leaks can be felt or heard. Small leaks can be detected by coating joints with soap-and-water solution. Bubbles indicate leaks. If a leak is revealed between

Fig 32. Testing pressure tank and valve, using 0-3,000-pound testing gage from service kit.

pressure tank and pressure-tank valve, or between check valve and pressure-tank valve, replace all three as a unit. If tests do not show up the leak the tank may have been improperly charged. It should be recharged and then retested.

54. SERVICE WHEN FIRING.

a. Failure to ignite. Pull trigger repeatedly. If ignition cylinder still fails to ignite, dirt may be wedged in ignition head. Unscrew shield one-half turn. Screw it back, rapping shield with the hand while turning. This should dislodge foreign matter. Pull trigger again. Repeat procedure, if necessary.

b. Safety head "blows" (breaks). If safety head breaks, firing mission cannot be carried out. On return, have head replaced. (Par 69) Follow test procedure. (Par 56 b)

55. SERVICE AFTER FIRING.

a. Unloading. Remove ignition cylinder (Par 30), close pressure-tank valve, and blow out remaining fuel and pressure. (Par 30)

b. Removal of equipment. Release the body straps, then the shoulder straps. If prone, lie on side and allow tank group to roll off onto ground. If standing or kneeling, take care that tank group does not drop on feet or legs.

c. Correcting or reporting. Correct any failures or difficulties or report them as soon as possible to service or maintenance personnel.

d. Gun. Remove shield (Par 18) and clean interior of shield with cloth. Clean holes in shield with wire or wooden splint. Clean external surfaces of barrel, nozzle, needle, and other parts. Check cleanliness and adjustment of needle. (Par 75 d) Check trigger for operation. Lubricate. (Par 49)

e. Fuel tanks and passages. Remove plugs (Par 69 b). Drain any remaining fuel. Use gasoline to remove residues of thickened fuels before they have a chance to harden and obstruct passages. If necessary, fill tanks with gasoline and allow to stand for several hours, shaking occasionally. Drain and repeat if necessary.

f. Safety head. Check head to see if it is ruptured; if it is, replace. (Par 69) Follow test procedure. (Par 56 b)

g. Pressure-tank valve. If weapon is to be stored, open pressure-tank valve and leave it open until next charging.

h. Carrier. Scrub, if necessary, with soap and water, or gasoline.

i. Exterior metal surfaces. Scrub exterior metal surfaces clean of fuel to prevent fire hazard. Allow to dry before using again.

j. General inspection. Carefully examine all other parts, adjust as necessary, and replace any which are damaged.

56. SERVICE AFTER SIX FIRING MISSIONS.

After the flame thrower has been used on six firing missions or the equivalent in training work, experienced personnel should follow these directions:

a. Before-operation and after-operation service. Follow the same procedures as in Paragraphs 52, 53, and 55.

b. Test firing (or simulated firing). (1) If tactical conditions permit test firing at a suitable test range (Par 15), fill the fuel tanks with fuel. (Pars 37 through 40)

(2) If test firing with fuel is impracticable, fill fuel tanks with clean water. (Be sure to dry all parts after test.)

(3) Remove filling-plug assembly. (Par 69) Fish out the retainer rod and chain by means of a bent wire.

(4) Do not unscrew the safety-head plug.

(5) Insert the testing plug with 0- to 500-pound pressure gage (plug and gage are from service kit) in the filling-plug opening Tighten testing plug in seat with wrench.

(6) Fully charge pressure tank. (Par 32)

(7) If test firing with fuel, load ignition cylinder. (Par 18)

(8) Open pressure-tank valve and simultaneously observe pressure in fuel tanks by reading gage. The gage indicates the pressure in both tanks. It should be between 350 and 390 pounds per square inch.

(9) Read the gage at the expiration of not less than 5 minutes. The tanks should have a pressure reading of not more than 390 pounds. If the pressure continues to increase beyond 390 pounds and the safety head blows, replace the safety head and the pressure regulator.

(10) Fire by operating controls (or simulate firing if tanks are filled with water). The burst should last 3 seconds, during which time the pressure should not drop below 260 pounds.

(11) If the pressure does not conform to the requirements stated in (8), (9), and (10), adjust the pressure regulator upward or downward. (Par 67 d)

(12) While the above test firing is proceeding, check for leaks at all joints and connections on the tank group. The pressure system should be checked by painting the joints with soap-and-water solution and by looking for bubbles which indicate leaks. For replacement of parts where pressure leaks, see Paragraph 66. Fuel leaks may be seen without soap and water. For repair of fuel leaks, see Paragraph 75 e. The nozzle should be observed by removing the ignition shield. Nozzle leaks are corrected by cleaning, adjusting needle (Par 75 d), or by regrinding. (Par 75 e) If this is not successful, replace both needle and barrel as a unit.

c. Fuel valve. Discharge all pressure from the gun by operating the fuel valve. Carefully remove the valve grip and grip support. (Par 74) Look for signs of leakage at the valve diaphragm. If a leak is present, replace valve-diaphragm assembly. (Pars 75 b and 75 c)

d. Valve grip. Disassemble the valve grip (Par 74) and lubricate. (Par 49)

e. Carrier. Tighten the carrier cord.

f. Gun interior. If thickened fuel has been fired, disassemble the gun. All parts should be cleaned of accumulations of dried fuel. Lubricate (Par 49) and reassemble. If liquid fuel has been fired, flush gun with clean gasoline. Disassemble only enough to lubricate. Reassemble.

Section XV
TROUBLE SHOOTING

57. PRECAUTIONS.
First, remove the ignition cylinder. Then, before disassembling, servicing, or repairing parts which may be under pressure, be sure to release the pressure. Remove fuel, when necessary.

58. FUEL LEAKS.

Trouble	Remedy
a. Defective or damaged valve-diaphragm assembly.	If leak is observed in valve grip, disassemble. (Par 74) If diaphragm is torn, or damaged in any other way, remove and replace. (Par 75)
b. Defective threaded connections on fuel lines.	Disconnect, using wrenches. If thread is stripped or badly damaged, replace the threaded part. If threads appear to be sound, clean them and reconnect. If leak is between tank coupling and tank connector or between hose and fuel-valve body, apply anti-seize compound before rescrewing. Tighten joint with wrenches.
c. Dirt or foreign matter on seats or threads.	Clean parts carefully with cloth before reassembling.
d. Leak at nozzle.	Adjust needle. (Par 75 d) If leak persists, either replace needle and barrel as a unit or use lapping compound on parts. Turn needle in seat until parts make a tight connection when seated. Remove lapping compound and reassemble.

e. Worn body of hose.   Replace fuel-hose assembly. (Par 73)

f. Leak at tank coupling.   Remove and replace coupling washer if damaged. (Par 70) If hose nipple, tank end, is damaged, repair nipple (Par 73 d) or replace fuel-hose assembly.

59. SAFETY HEAD "BLOWS" (BREAKS).

Trouble	Remedy
a. Defective safety head.	Replace with new safety head. (Par 69 b)
b. Defective pressure regulator.	If replacement safety head also breaks, follow test procedure in Paragraph 56 b to determine whether pressure regulator needs adjustment or is defective.

60. CARRIER UNCOMFORTABLE.

Trouble	Remedy
a. Cord becomes loose or breaks.	Use only hard-braided seine cord furnished in service kit for replacements. Lace tightly as shown in Figure 46, using slip-proof knots at ends.
b. Straps not adjusted to fit wearer.	Adjust straps to fit each new wearer. Tank group must be high on back and snug on body. (Pars 19 and 71)
c. Carrier frame presses on wearer's back.	Cord is too loose. Tighten cord. Use slip-proof knots at ends.

61. SHORT RANGE.

Trouble	Remedy
a. Stream of burning fuel issues at an angle or in a very broad spray.	Fuel valve is not fully open because of:
	(1) Faulty operation. Be sure to compress controls all the way when firing. (Par 26)
	(2) Improper adjustment or assembly of valve. To correct, see Paragraphs 74 and 75.
b. Rapid drop of range during a burst.	Pressure-tank valve is not fully open. Open all the way. If this is not effective, test pressure regulator. (Par 67 d)
c. Shorter range in each successive burst.	Pressure tank is not fully charged.
	(1) Before firing be sure tank is

charged to at least 1,700 pounds per square inch. (Par 32)

(2) Check for leaks to make sure pressure has not decreased since charging. (Par 53 d)

d. Short range with longer time of discharge than 8 to 9 seconds.

Dried fuel or other foreign matter is in fuel lines. Disassemble and clean.

62. FUEL-VALVE FAILURE.

Trouble

Valve fails to close when controls are released.

Remedy

(1) Work the grip safety to trip the valve lever.

(2) Foreign matter may be in barrel, or barrel may be dented. If dented, replace barrel and needle as a unit. If not dented, disassemble and clean. (Pars 74 and 75)

63. FAILURE OF IGNITION CYLINDER TO IGNITE.

Trouble

a. Match in cylinder moves but incendiary charge does not ignite.

Remedy

Pull trigger repeatedly. If cylinder does not ignite, remove cylinder (Par 30) and examine.

(1) If matches have been pushed flush with inner surface of cylinder body, the cylinder is defective. Destroy. (Par 30) Replace.

(2) If matches project 1/16 inch or more from cylinder, ignition head is defective. Disassemble ignition head (Par 76 b) and examine. Replace parts as necessary. (Par 76 c)

b. Cylinder does not rotate to bring new charge into position.

(1) Spring case is not free to rotate because of dirt. Clean and lubricate. (Par 49)

(2) Cylinder is improperly loaded. (Par 18)

(3) Ignition cylinder binds on barrel because of dirt or excessive warping of ignition cylinders from heat of firing. Remove and destroy (Par 30) ignition cylinder. Reload.

(4) Spring case is defective. Re-

c. Trigger does not re-
turn to normal position
(with ignition cylinder in
place).

place as a unit. (Par 76 b, c)
(1) When on a mission, use fingers
on trigger to pull back to normal position.
(2) If time permits, remove trigger rod. (Par 76 b) Clean rod
and hole in which rod slides.
Lubricate. (Par 49) Reas-
semble. (Par 76 c)

d. Lack of spring tension
at trigger.

Trigger spring is off hook of trigger, off spring screw, or broken.
Replace where necessary.

64. FAILURE OF FUEL TO IGNITE.

Trouble

Remedy

a. Atomizer hole clog-
ged.

Clean with fine wire. (Par 52 i)

b. Fuel troubles at low
temperature.

(1) At temperatures below minus
20 degrees Fahrenheit, ignition
of any standard fuel is uncer-
tain. Operation at these temperatures should be avoided
unless tests of fuels by firing
with flame throwers are first
made.
(2) At temperatures above minus
20 degrees Fahrenheit, no difficulty should be experienced
with thickened gasoline. When
blended fuels are used, the
ratio of gasoline content should
be increased as temperature
decreases.

c. Failure of ignition cylinder.

See Paragraph 63.

Section XVI
TANK GROUP

65. GENERAL.

The tank group stores fuel and pressure. The fuel is placed under pressure when the pressure-tank valve is opened. The tank
group is supported upon the firer's back and shoulders by the
carrier.

66. PRESSURE TANK AND VALVE ASSEMBLY.

a. Description and functioning. The pressure tank and valve assembly (Fig 33) includes:

(1) Pressure tank. The pressure tank is a lightweight, airplane-type cylinder, able to withstand the great pressure which it contains. The tank is charged with air or nitrogen at 1,700 to 2,100 pounds per square inch pressure by use of auxiliary equipment as described in Paragraphs 31 and 32. This pressure stays in the pressure tank until the weapon is ready to be fired. Opening of the pressure-tank valve releases air or nitrogen through the pressure regulator to the fuel tanks. Oxygen or combustible gases are never used in the tank because a violent explosion may result. The tank is large in capacity to assure ample pressure, and hence full range, for the entire load of fuel   The pressure-tank clamp (Fig 39), a steel-strap device with hinge and toggle-type latch, holds the pressure tank in place on the fuel tanks.

(2) Pressure-tank valve. (Figs 33 and 34) This valve is screwed into the bottom of the pressure tank. The valve stem slides into the valve end of the valve flexible shaft. When opened by means of the valve handle and valve flexible shaft, the valve permits passage of compressed air or nitrogen through tubes and the pressure regulator to the fuel tanks. The valve is of the quick-opening, packless, diaphragm type.

(3) Pressure-valve handle and valve flexible shaft. (Figs 33 and 34) The pressure-valve handle is held by a small nut on the end of the valve flexible shaft, which in turn is connected to the pressure-tank valve by means of the valve stem and a large hex nut. The handle and shaft extend to the right of the tank group, enabling the firer to open and close the valve without assistance when carrying the weapon. The handle slips over the end of the shaft and is held to it by a nut. The shaft is held to one of the fuel tanks by a clamp, nut, and bolt welded to the tank.

(4) Check valve. (Figs 33 through 35) The check valve has the same function as the valve on a vehicular tire tube, but it is much heavier in construction and different in design because the pressure in the flame thrower is 50 times greater than that in an automobile tire tube. Connected by threads to the pressure valve, the check valve permits compressed air or nitrogen to enter the pressure tank during charging (Pars 31 and 32) but prevents its escape when the outside source of pressure is removed. The cap is removed from the check valve only for charging or testing.

b. Removal. (Fig 33) To prevent damage to threads, leaks, and loss of pressure and range, remove pressure tank and valve assembly only when necessary.

(1) Release of pressure. Be sure all pressure has been released from the pressure system before disassembling or removing any part or assembly of the pressure system. To release pressure, operate fuel valve (Par 26) and hold open until

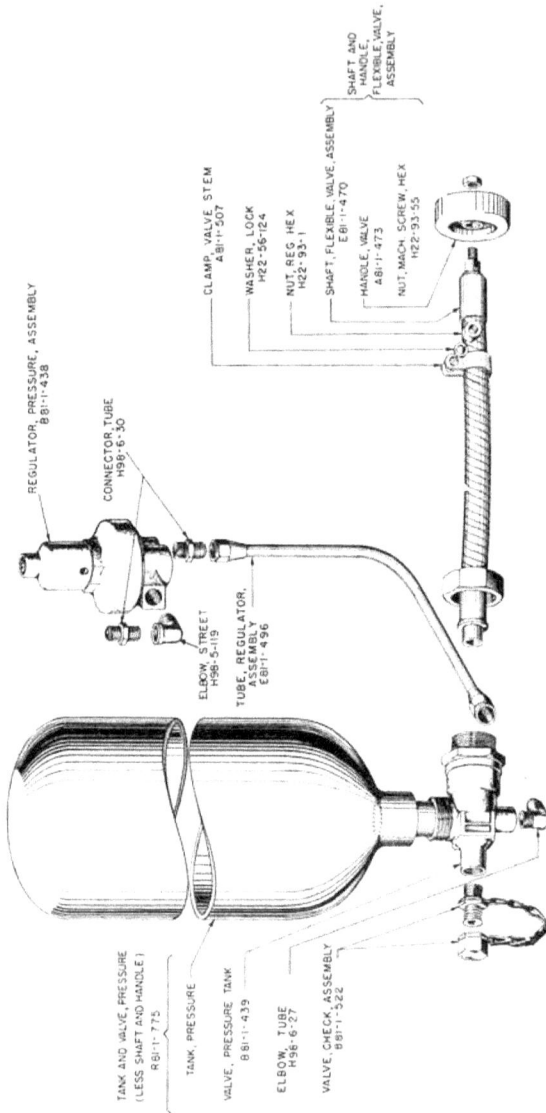

Fig 33. Pressure system disassembled, showing nomenclature and Chemical Warfare Service stock numbers for requisitioning spare parts.

pressure is exhausted. As an added precaution, personnel should avoid facing the connections when disconnecting parts or assemblies.

(2) Removal procedure. After release of all pressure:

(a) Loosen the clamp from the valve flexible shaft.

(b) Using wrench, unscrew the large hex nut which holds the flexible shaft on the pressure-tank valve.

(c) Pull the valve flexible shaft and handle free of the valve.

(d) Using wrench, unscrew the flared tube nut on the regulator tube adjacent to the pressure-tank valve.

(e) Open pressure-tank clamp (Fig 39) and swing clamp strap outward.

(f) Remove the pressure tank together with the pressure-tank valve and check valve.

(g) To remove valve handle, use the adjustable-end wrench to loosen and remove nut from threaded outer end of valve flexible shaft. Slide out the valve handle.

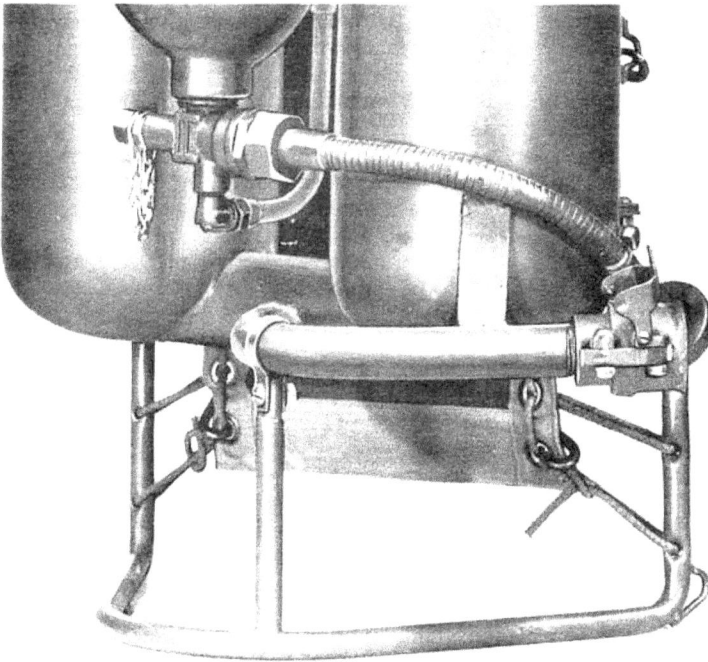

Fig 34. Lower portion of pressure system, assembled.

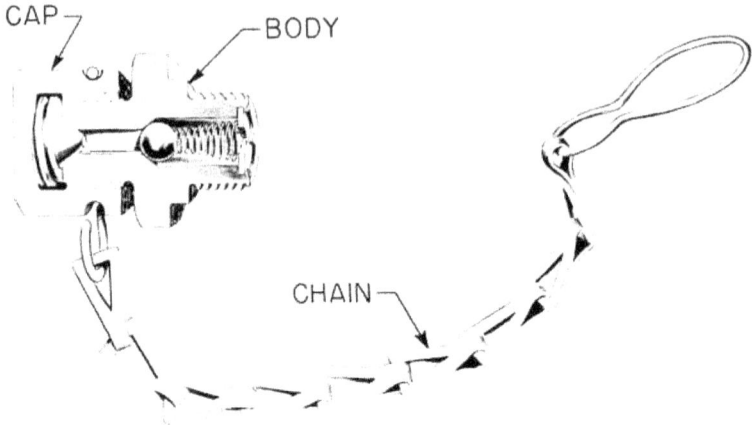

Fig 35. Check valve (cross section).

c. Installation. (Figs 33 and 39) To install:
(1) Insert pressure tank (with pressure-tank valve and check valve mounted on the tank) through the pressure-tank clamp. Be sure to aline the regulator tube, elbow, and pressure-tank valve threads carefully so that they cannot be damaged when connecting.
(2) Close the pressure-tank clamp.
(3) Start the threaded connections by hand to be sure they are well alined. Do not force. Use wrench for final tightening, but do not apply great torque to the wrench.
(4) Insert the valve flexible shaft through small clamp into the pressure-tank valve. Using wrench, tighten the large hex nut located between shaft and valve.
(5) Tighten the clamp on the valve flexible shaft.
(6) Place pressure-valve handle on threaded end of the shaft. Place nut on threaded end and tighten with adjustable-end wrench.
d. Adjustment. If valve handle cannot be turned by hand:
(1) Remove flexible shaft and handle. Never apply a wrench to these parts.
(2) Turn end of pressure-valve stem with wrench to open valve.
(3) If stem will not turn, replace the tank and valve.
(4) If stem turns, work it back and forth with wrench.
(5) Reconnect flexible shaft and handle.
(6) If handle does not turn easily, repeat the process until handle turns, or replace tank and valve.
(7) Close valve before charging tank.
e. Maintenance.
(1) If either pressure tank, pressure-tank valve, or check valve are damaged or defective, all three must be replaced as a unit. No attempt may be made to repair any of these parts or

their connections. If makeshift repairs or improvised parts are devised, serious accidents can result because of the extremely high pressures to which the equipment is subjected.

(2) Keep all threaded connections tightened. If a leak is suspected at any threaded connection, follow procedure in Paragraph 53 d.

67. PRESSURE REGULATOR.

a. Description and functioning. The regulator automatically reduces the variable pressure of air or nitrogen in the pressure tank to a constant operating pressure of approximately 350 pounds per square inch in the fuel tanks. The regulator is located at a protected position in the tank group of the M2-2 portable flame thrower, where it is not readily subject to tampering or damage from the outside. The regulator tube with fittings connects the pressure-tank valve and pressure regulator. (Fig 33) Its outlet is connected to the fuel tanks by the diffusion-pipe assembly. (Par 68 a) Either of two interchangeable types of regulators is furnished: the spring type (Figs 33, 36, and 37), and the dome type (Fig 38).

b. Removal of pressure regulator. After release of all pressure:

(1) Remove carrier (Par 71 b), if necessary.

(2) Using wrench, unscrew the flared tube nuts and other fittings.

(3) Lift out the pressure regulator.

c. Installation of pressure regulator. Line up pressure regulator, regulator tube, diffusion-pipe assembly, and fittings carefully so they will not be damaged when threads are tightened. Start threads with the hands. Apply only moderate wrench pressure to complete tightening. Replace carrier or carrier pack if either has been removed.

d. Adjustment of pressure regulator. The pressure regulator ordinarily requires no attention other than checking and tightening connections with the regulator tube and the diffu-

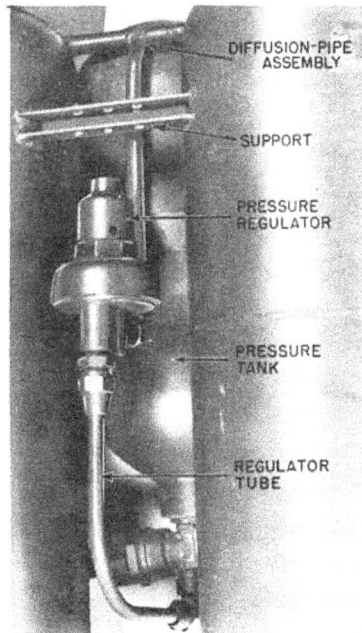

Fig 36. Rear of tank group, with carrier removed to show pressure regulator (spring-type) and connections.

sion-pipe assembly. If a defect in the regulator is indicated by falling off of the range of the weapon or by frequent breakage of the safety-head diaphragm (Par 56 b), the following procedure should be carried out. (When using wrenches, do not apply excessive force.)

(1) Remove the filling plug (Par 69 b) and ignition cylinder. (Par 30 a)

(2) Fill the fuel tanks with 4 gallons of water (or fuel).

(3) Connect the 0- to 500-pound fuel-tank testing gage, found in the service kit, to the filling-plug hole. Tighten plug of gage with wrench.

(4) Charge the pressure tank to a pressure of 1,800 pounds per square inch. (Pars 32 and 33)

(5) Open the pressure-tank valve.

(6) Read the pressure on the gage. If 350 to 390 pounds is indicated, omit steps (7) through (10).

(7) To increase the pressure of a spring-type regulator:

(a) Pry off the protective cap.

(b) Turn a set-screw wrench clockwise in the adjusting screw and read pressure on gage.

(8) To decrease the pressure of a spring-type regulator:

(a) Turn the set-screw wrench counterclockwise more than is considered sufficient to effect the desired reduction.

(b) Turn the pressure-tank valve off.

(c) Relieve pressure in the fuel tanks by compressing the fuel valve until the pressure is below that desired.

(d) Release the fuel valve.

(e) Open the pressure-tank valve and allow the system to reach a state of equilibrium, which occurs when the hissing sound ceases.

(f) Repeat the steps described above in (6) and (7).

(9) To increase the pressure of a dome-type regulator:

(a) Open needle valve No. 1 one full turn. (Fig 38)

(b) Open needle valve No. 2 one full turn. (There will be slight leakage around the needle-valve thread.)

(c) Open needle valve No. 3 very slowly, watching pressure gage closely. (As pressure builds up in fuel tanks there will be slight leakage through needle valve No. 1.)

(d) When the pressure gage indicates 350 pounds, close needle valve No. 3 tight.

(e) Close needle valve No. 2 tight.

(f) Close pressure-tank valve.

(g) When gage indicates zero, close needle valve No. 1 tight.

(10) To decrease the pressure of a dome-type regulator:

(a) Open needle valve No. 1 one full turn. (Fig 38)

(b) Open needle valve No. 3 very slightly, which will lower the pressure.

(c) When 350 pounds is reached, close valve No. 3 tight.

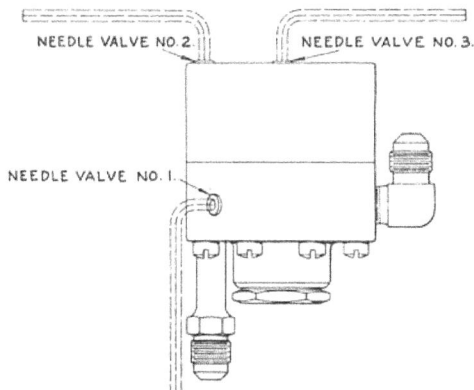

Fig 37. Pressure reg- Fig 38. Pressure regulator, dome-type,
ulator, spring-type.   showing needle valves and wrenches.

(d) Close pressure-tank valve.

(e) When gage indicates zero, close needle valve No. 1
tight.

(11) Open pressure-tank valve and press the fuel valve to observe the pressure with the weapon operating.

(12) After final adjustment:

(a) Close the pressure-tank valve.

(b) Open the fuel valve and release the pressure from the
fuel tanks.

(c) Remove the pressure gage and plug from the fuel tank.

(d) Install filling plug.

(e) Tighten filling plug with wrench.

(f) If regulator is spring-type, replace its protective cap.

68. FUEL-TANK ASSEMBLY.

a. Description and functioning. (Figs 4, 5, and 39) The fuel
-tank assembly includes:

(1) Fuel tanks. Two alloy steel fuel tanks hold the fuel before
it is propelled to the target. They have a combined capacity,
including void, of 4-1/2 gallons. A void of approximately 1/2
gallon is left in tanks when filling to allow for expansion and to
permit entry of the compressed nitrogen or air. To speed filling
and cleaning of the tanks, two openings are provided on top of the
fuel tanks. The openings are threaded to receive the filling-plug
assembly and the safety-head plug assembly, which are interchangeable in the openings. Filling operations involve the use of
auxiliary equipment and are described in Paragraphs 34 through
40. The carrier and the pressure system are supported on the
fuel tanks.

(2) Tank connector. This open passageway between the fuel
tanks makes them, in effect, a single container. The location of the

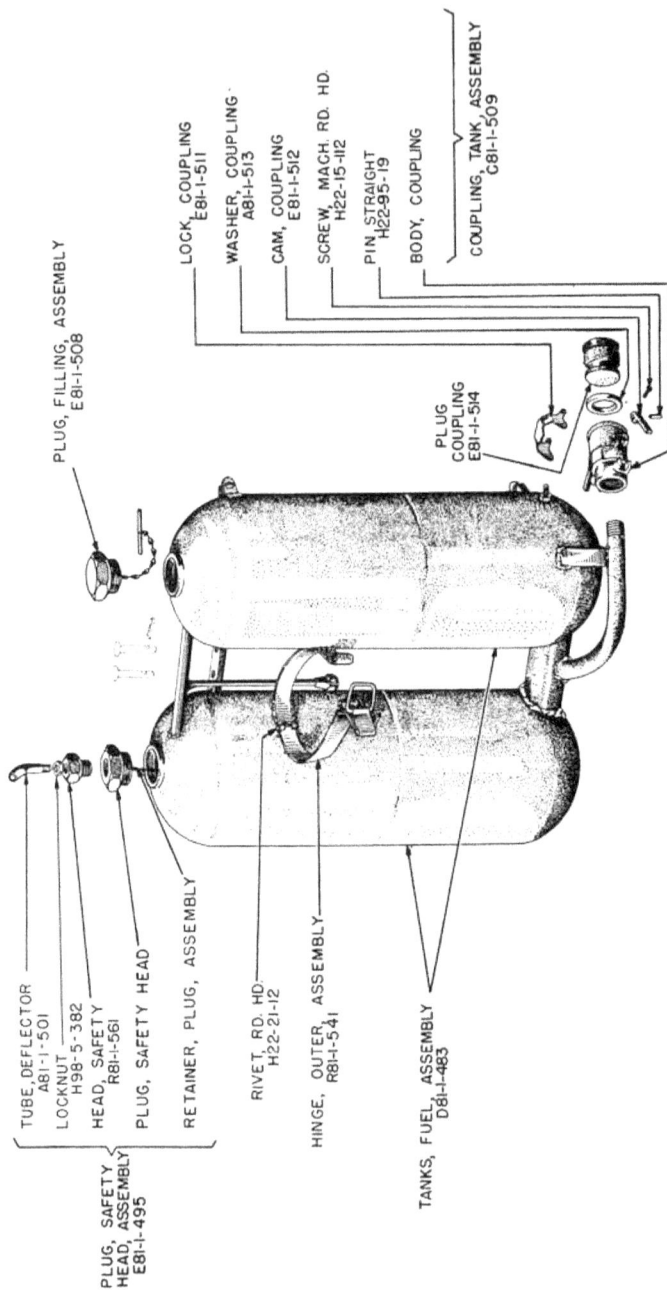

Fig 39. Fuel system of tank group and related parts disassembled, showing nomenclature and Chemical Warfare Service stock numbers for requisitioning spare parts.

tank connector and its large diameter permit easy flow of fuel and pressure between the two tanks.

(3) Hose connector. The hose connector is the outlet for fuel from the fuel tanks. It is located so that nearly all the fuel is propelled from the weapon if firing positions are correct. (Par 24) One end is welded to an opening in the tank connector. The other end is threaded into the tank coupling.

(4) Frame clamp. This small metal clamp, with bolt, nut, and washer, holds the hose connector to the carrier frame.

(5) Diffusion-pipe assembly. This T-shaped tubing carries compressed air or nitrogen from the pressure regulator to each of the fuel tanks. A flared tube connection and elbow connect the stem of the T to the pressure regulator. The horizontal tubes of the T extend into the fuel tanks and are welded to the fuel tank walls. Within the fuel tanks these tubes are perforated with holes which permit ready escape of the compressed nitrogen or air into the fuel tanks when the pressure-tank valve is open.

b. Removal and installation. The tank connector, hose connector, diffusion-pipe assembly, and the two fuel tanks are welded together and cannot be disassembled from each other. No attempt should be made to remove any of these parts or assemblies.

c. Maintenance. Other than cleaning (Pars 51 d and 55 e), repainting, and tightening of threaded joints, no repairs will be attempted by the first or second echelon on the fuel tanks, tank and hose connectors, or diffusion-pipe assembly. Emergency repairs may be made only by the third or fourth echelon. No attempt should be made to weld or patch any part of the fuel tanks.

69. FILLING AND SAFETY-HEAD PLUG ASSEMBLIES.

a. Description and functioning.

(1) Filling-plug assembly. (Fig 39) This assembly fits into the 1-3/8-inch threaded opening at the top of either one of the fuel tanks. It permits filling and cleaning of the tanks, and seals the opening when the tank is not being filled or cleaned. The assembly includes the filling plug proper and a plug-retainer assembly. The latter is a metal rod which hangs from the plug on a metal chain. The rod and chain prevent accidental loss of the plug.

(2) Safety-head plug assembly. (Figs 39 and 40) This assembly is screwed into the threaded opening on top of either fuel tank. It serves the same functions as the filling-plug assembly and moreover protects the firer and other personnel. It includes:

(a) Safety-head plug. This plug is similar to the filling plug except for the threaded hole which receives the safety head.

(b) Safety head. This metal head screws into the safety-head plug. It includes a soft metal diaphragm which bursts when the pressure in the fuel tanks exceeds 500 pounds per square inch.

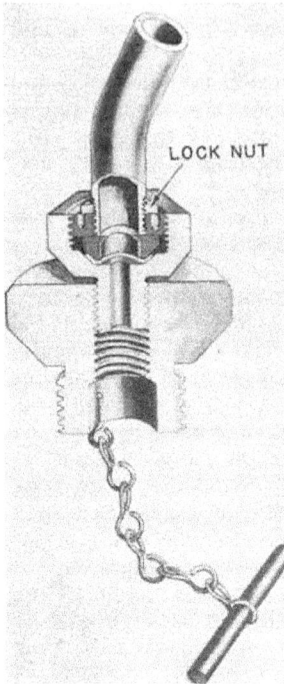

Fig 40. Safety-head
plug assembly
(cross section).

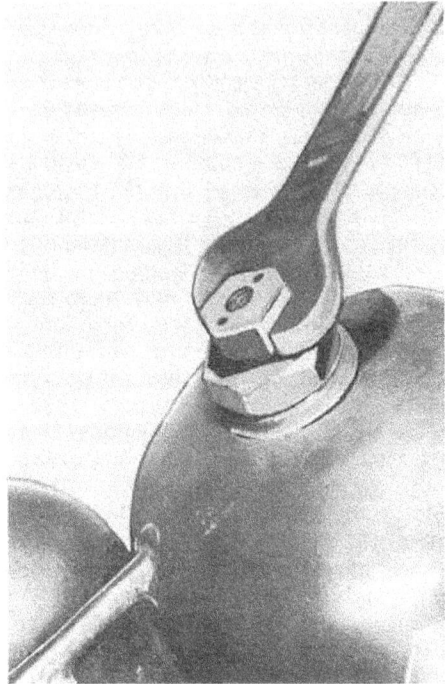

Fig 41. Unscrewing safety head
from safety-head plug,
using wrench.

It prevents the building up of dangerous pressures in the fuel tanks.

(c) Deflector tube. This short, curved piece of 1/8-inch pipe deflects fuel and pressure away from the firer if the safety head bursts. A lock nut holds the tube in position. (Par 12 j)

(d) Plug-retainer assembly. This assembly consists of a metal rod and chain which hang from the plug and prevent accidental loss of the plug when filling or inspecting.

b. Removal of plugs. (1) Before removing the filling plug, the safety-head plug, or an unbroken safety head, operate the fuel valve until any pressure which may have accumulated in the fuel tanks is eliminated. If the coupling plug is in the tank coupling, very slightly loosen the threads of either the filling plug or the safety-head plug, using the 1-3/4-inch wrench, to eliminate pressure in the fuel tanks. Keep face and eyes away from the threads.

(2) The plug-retainer assemblies should not be lifted completely out of the tanks unless required.

(3) If either the rod or the rod and chain breaks from one of the plugs and falls into the tank, upend the tank group to permit removal of the parts.

(4) To replace burst safety head, unscrew lock nut and deflector tube. (Fig 11) Using wrench (Fig 41), unscrew safety head. Never disassemble the safety head.

c. Installation of plugs. The filling plug, safety-head plug, and safety head are screwed in by hand and then tightened with wrenches. No substitution will be made for the safety head, which is manufactured to burst at the safe limit of pressure. The plug threads and seats should be cleaned with a cloth (Fig 29) before installing plugs. Screw in deflector tube, using hand pressure. The tube outlet should face to the rear and at a 45-degree angle to operator's left shoulder. (Fig 18) Replace lock nut and tighten with wrench. (Use wrench on lock nut, not on deflector tube.)

d. Maintenance of plugs. Replace safety head if damaged or blown. Never repair safety head or use an improvised head.

70. TANK COUPLING.

a. Description and functioning. This quick-connecting coupling (Fig 42) connects and locks the fuel hose or the coupling plug to the tank group. The coupling cams, lock, and washer provide a secure and tight joint. The tank coupling makes possible rapid replacement of emptied tank groups with filled and charged tank groups in the field. No tools are needed for this operation.

b. Removal. (1) To remove the tank coupling from the hose connector, apply a wrench and unscrew.

(2) To disconnect the tank coupling from the fuel hose or the coupling plug:

(a) Release pressure from fuel tanks by operating the fuel

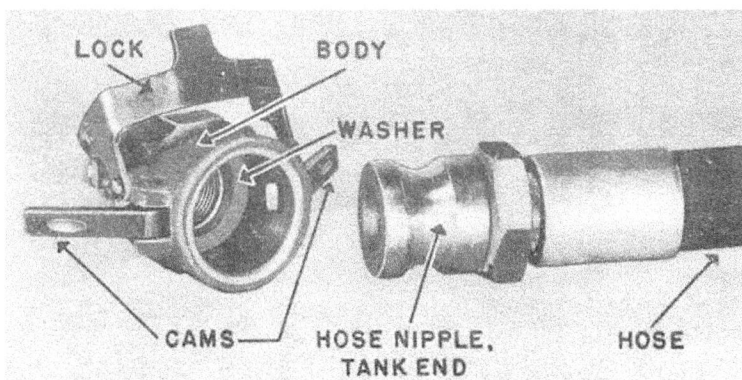

Fig 42. Tank coupling and end of fuel-hose assembly.

69

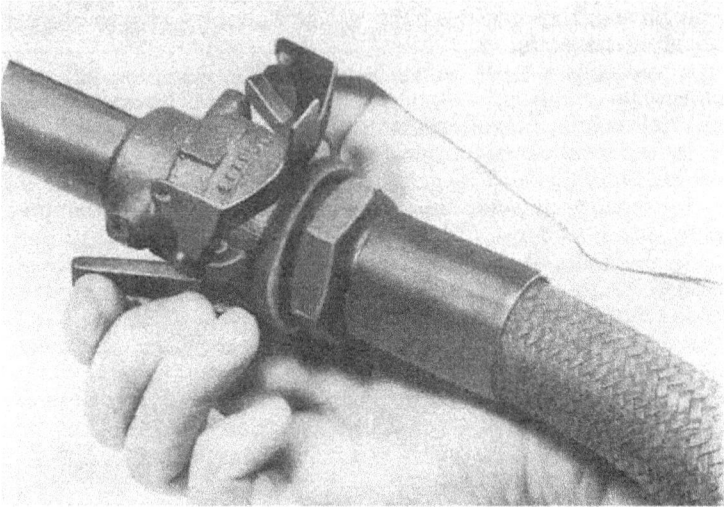

Fig 43. Closing cams of tank coupling to connect gun and tank group. This is done before locking. (See below.)

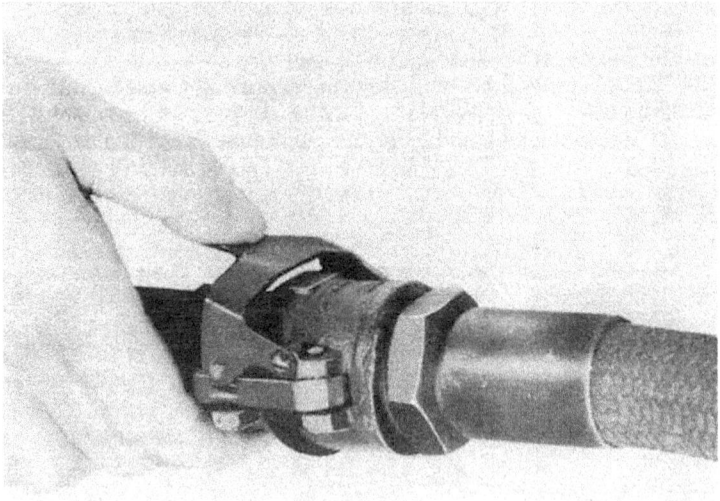

Fig 44. Closing lock of tank coupling to secure gun to tank group. This also provides a fuel-tight seal.

valve or by opening very slightly the filling plug.

(b) Using hands, pivot
the coupling lock back on the
coupling body.

(c) Using hands, pivot
the two coupling cams back on
the coupling.

(d) Slide out the fuel hose
or the tank coupling.

(e) If coupling washer is
to be removed, pry out with a
screw driver.

c. Installation of tank coupling. Proceed as follows:

(1) If coupling washer has
been removed, replace.

(2) Insert coupling plug or
hose nipple, tank end, in the
coupling as far as it will reach.
Close the two cams. (Fig 43)

(3) Close the coupling lock

Fig 45. Coupling plug in place
in tank coupling. This arrangement is used when fuel tanks
are brought back for filling
with gun detached.

(Fig 44), being sure to push it all the way, until it covers the ends
of both cams. (Figure 45 shows coupling lock correctly locked
on coupling plug.)

(4) If the tank coupling has been removed from the hose connector, screw it on hand tight. Anti-seize compound should be applied lightly to the threads to assure a tight joint. Use wrench
to tighten the coupling until it is in the position shown in Figure 34.

d. Maintenance of tank coupling. The coupling washer, made
of synthetic rubber, should be inspected frequently. If it is damaged or swollen, remove it and replace. If the coupling leaks,
inspect, and if necessary, remove and replace the washer.

71. CARRIER.

a. Description and functioning. (Fig 46) The tank group is
securely carried on the firer's back and chest by the carrier,
which includes the metal carrier frame, the canvas carrier
pack, webbing straps, and cord, all of which are parts of the
tank group.

(1) Carrier frame. This lightweight, tubular-metal frame
is bolted to two pairs of brackets (upper and lower) on the fuel
tanks. It is also bolted to the hose connector by the frame clamp,
which helps support the connector. The frame is pierced by two
parallel series of holes, through which the cord (lashing) of the
carrier is laced.

(2) Carrier pack. This is a sheet of heavy canvas, reinforced
on the tank side with strips of webbing. The smooth side of the

carrier pack rests against the firer's back and cushions the back from contact with the metal tanks. A series of eyelets is located on each side of the pack.

(3) Seine cord (lashing). The carrier pack is fastened to the carrier frame by means of hard-braided cord which is laced through the eyelets in the pack and the holes in the frame. The cord that comes on the flame thrower stretches very little under load.

(4) Straps. The straps, made of wide cotton webbing, are adjustable to fit the wearer. (Fig 18) They are provided with snap release, hook-and-eye, and snap fasteners. The shoulder straps have quick-release fasteners for rapid removal, if necessary, of the tank group from the firer. The upper ends (steel loops) of the shoulder straps are secured by pins to the steel support which connects the two fuel tanks. Each of the pins is held in position by a split cotter pin, which is inserted through a hole in the pin and is then spread. The lower ends of the shoulder straps snap onto metal loops at the bottom of the carrier frame. The upper body straps are attached to metal loops on each side of the carrier frame. The lower body straps are fastened to one of the lower two pairs of eyelets of the carrier pack.

b. Removal of carrier. (1) To remove carrier or carrier frame, use screw driver and adjustable-end wrench to take off frame clamp, bolt, nut, and lock washer. (Fig 34) Then remove two pairs of bolts, nuts, and lock washers which hold the carrier frame to bottom and top of the fuel tanks. Lift off the carrier.

(2) To remove body straps, unsnap ends and lift out of holes. To remove shoulder straps, unsnap lower ends and remove from holes. Pull out cotter pins, then pins, from upper ends of shoulder straps, and lift out straps.

(3) To remove carrier pack, unknot and unlace cord.

c. Installation of carrier. (1) To install carrier frame (or a complete carrier) place frame in position adjacent to fuel tanks (Fig 46), insert bolts in holes, place lock washers and nuts on bolts, and tighten with screw driver and wrench. Replace frame clamp on fuel connector and frame. Insert bolt in holes, place lock washer and nut on bolt. Tighten with screw driver and wrench.

(2) If carrier pack has been removed, use cord to relash. Lace tight and use slip-proof knots. (Fig 46)

(3) To install straps, snap ends of body straps and lower ends of shoulder straps into positions shown in Figure 46. Place upper ends (steel loops) of shoulder straps in steel support between fuel tanks. Insert two pins through any two of the holes in the support and through the shoulder strap loops. Insert cotter pins in holes in pins and spread cotter pins to lock pins in place.

d. Adjustment of carrier. Carrier must be carefully adjusted to fit the individual firer so the load will not shift during sudden, rapid changes of firer's position. Adjustments are as follows:

(1) Cord and carrier pack. Cord must be tight at all times. The cord furnished with the equipment has very little tendency to stretch. However, pull cord tight when lacing and use slip-proof knots at ends. Tighten cord periodically.

(2) Straps. Adjust straps to fit each firer, moving slides on straps as necessary. Straps must fit snugly to prevent shifting of load and to keep tank group high on firer's back. Lower body straps may be fastened in second pair from the bottom of carrier

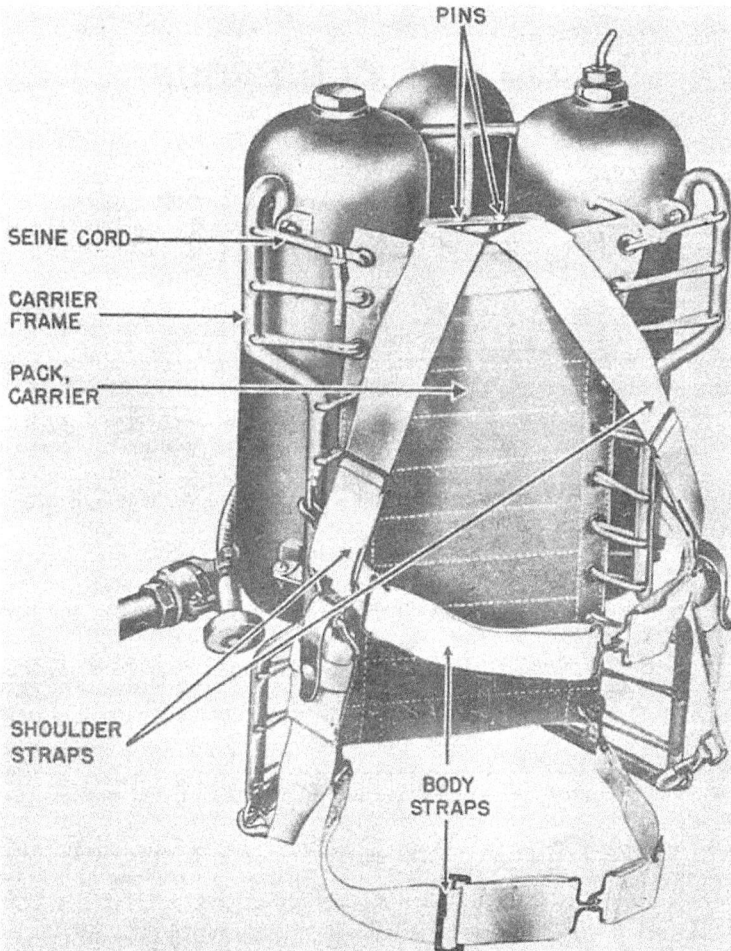

Fig 46. Carrier assembled on tank group.

eyelets to conform to firer's physique. Pins, which hold top ends of shoulder straps to steel support between the fuel tanks, may be moved to any two of the three holes so as to provide the best balance in the load.

e. Maintenance of carrier. Keep carrier dry and clean. If flame thrower becomes wet or muddy, clean and dry carrier thoroughly. Store in a dry place. If rotted, mildewed, or damaged, replace affected parts. If cord frays or breaks, use special seine cord from service kit as replacement.

Section XVII GUN GROUP

72. GENERAL.

The gun group consists of the fuel-hose assembly and the gun. The gun includes the fuel valve, which controls the ejection of fuel, and the ignition head, which ignites the fuel.

73. FUEL-HOSE ASSEMBLY.

a. Description and functioning. (Fig 47) Hose, fuel, flame thrower, M1, assembly, provides a flexible connection between the fuel tanks and the gun.

(1) Hose. Made of synthetic rubber and reinforced with a cover of metal wire and cotton braid, the hose resists the action of gasoline and oil, and withstands a pressure of approximately 1,000 pounds per square inch. Its inside diameter is 7/8 inch; its outside diameter is approximately 1-1/4 inches.

(2) Nipples. The hose nipple, tank end, connects the hose to the tank coupling on the tank group. The hose nipple, gun end, is a threaded connector between the other end of the hose and the fuel-valve body.

b. Removal of fuel-hose assembly. Remove the hose from the gun only when necessary for maintenance. The threads in the fuel-valve body will be damaged by frequent screwing and unscrewing of the hose because the body is a lightweight aluminum casting. Fuel hose is replaced as a unit and is not disassembled in the second echelon. To disconnect from tank group, see Paragraph 70 b.

c. Installation of fuel-hose assembly. (1) To install in tank group, see Paragraph 70 c.

(2) To install in gun, apply anti-seize compound (from service kit) lightly to threads and screw hose in fuel-valve body by hand. Use wrench only enough to make a secure connection.

d. Maintenance of fuel-hose assembly. If the hose nipple, tank end, is badly nicked and does not provide a tight connection with a new coupling washer (Par 70):

(1) File the end surface, being careful to keep the surface at a right angle to the sides of the nipple.

(2) Couple hose nipple, tank end, to tank coupling. If coupling closes very easily, indicating washer is not being compressed, replace the washer and recouple. If coupling still closes too freely, the nipple has been filed too short, and the fuel-hose assembly should be replaced as a unit.

74. VALVE GRIP.

a. Description and functioning. (Fig 47) The valve grip is part of the fuel valve. It includes the controls and is held by the firer in his right hand to support the gun group. Parts of the valve grip are:

(1) Left and right valve grips. A pistol-type grip is formed by two aluminum housings designated as the left valve grip and the right valve grip. The two parts are held together by four screws and four lock washers.

(2) Grip support. This aluminum housing is mounted above the left and right valve grips and connected to them by two screws and lock washers.

(3) Valve lever. This control is made to fit the fingers and is mounted in front of and between the two parts of the valve grip. A pin at the top of the lever fits into holes in left and right valve grips and serves as a pivot, governing the movement of the lever. When the lever and the grip safety are compressed simultaneously by the operator, the valve is thereby opened and fuel is ejected from the gun.

(4) Grip safety. This control is grasped by the hand simultaneously with the valve lever. It is mounted back of and between left and right valve grips. A pin at the base of the safety fits into holes in the left and right valve grips and serves as a pivot in a manner similar to the pin on the valve lever. The fuel cannot be discharged unless both the valve lever and the grip safety are compressed simultaneously.

(5) Rocker arm. The rocker arm, a boat-shaped metal part, is mounted near its center on a pin. It is held in contact with the valve lever by means of a valve-grip spring and spring pin. At its top end, the rocker arm touches the yoke shaft of the valve-diaphragm assembly. When the valve lever and the grip safety are compressed, the rocker arm pushes the valve diaphragm assembly forward.

(6) Valve-grip spring. When the firer's hand releases the valve grip, the valve-grip spring forces the valve lever, the grip safety, and the rocker arm back to their normal, nonoperating positions.

b. Removal of valve grip. (1) Unscrew the four screws and lock washers that hold the grip support to the valve body. Remove the valve grip as a unit.

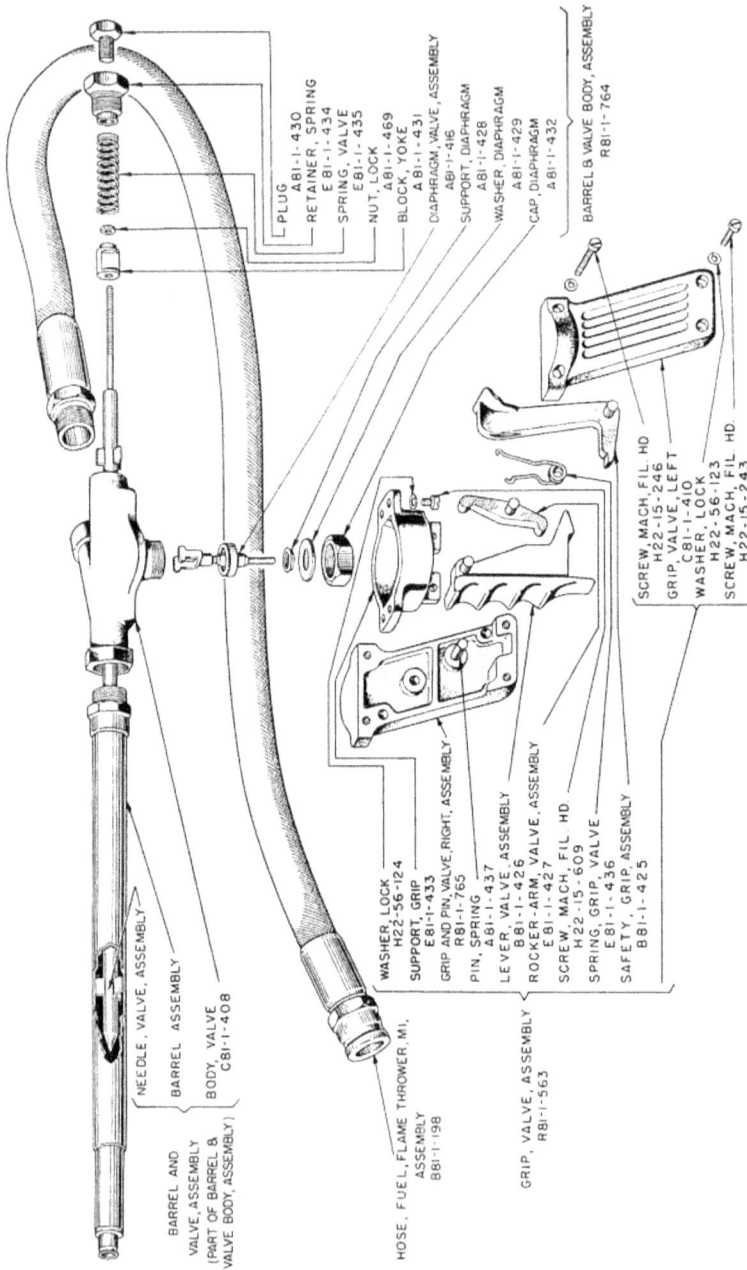

PLUG
A81-1-430
RETAINER, SPRING
E81-1-434
SPRING, VALVE
E81-1-435
NUT, LOCK
A81-1-469
BLOCK, YOKE
A81-1-431
DIAPHRAGM, VALVE, ASSEMBLY
A81-1-416
SUPPORT, DIAPHRAGM
A81-1-428
WASHER, DIAPHRAGM
A81-1-429
CAP, DIAPHRAGM
A81-1-432
BARREL & VALVE BODY, ASSEMBLY
R81-1-764

SCREW, MACH, FIL HD
H22-15-246
GRIP, VALVE, LEFT
C81-1-410
WASHER, LOCK
H22-56-123
SCREW, MACH, FIL HD
H22-15-243

NEEDLE, VALVE, ASSEMBLY
BARREL ASSEMBLY
BODY, VALVE
C81-1-408
BARREL AND
VALVE, ASSEMBLY
(PART OF BARREL &
VALVE BODY, ASSEMBLY)

HOSE, FUEL, FLAME THROWER, M1,
ASSEMBLY
B81-1-198

WASHER, LOCK
H22-56-124
SUPPORT, GRIP
E81-1-433
GRIP AND PIN, VALVE, RIGHT, ASSEMBLY
R81-1-765
PIN, SPRING
A81-1-437
LEVER, VALVE, ASSEMBLY
B81-1-426
ROCKER-ARM, VALVE, ASSEMBLY
E81-1-427
SCREW, MACH, FIL HD
H22-15-609
SPRING, GRIP VALVE
E81-1-436
SAFETY, GRIP, ASSEMBLY
B81-1-425

GRIP, VALVE, ASSEMBLY
R81-1-563

Fig 47. Fuel valve (disassembled) and fuel hose, showing nomenclature and Chemical Warfare Service stock numbers for requisitioning spare parts.

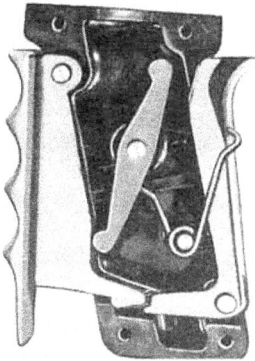

Fig 48. Location of parts in right valve grip before covering them with left valve grip.

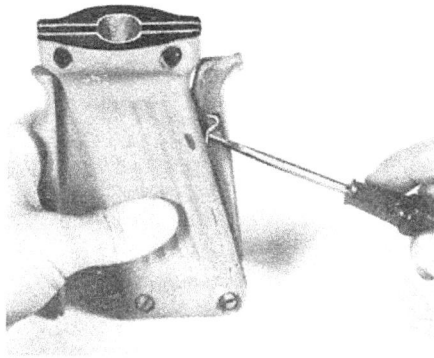

Fig 49. Using screw driver to push long end of valve-grip spring into groove in grip safety.

(2) To disassemble valve grip, remove screws and lock washers from the grip. Lift off the left valve grip, exposing contents of grip. Lift out the following parts: valve-grip spring, rocker arm, grip safety, and valve lever.

c. Installing valve grip. (1) Place grip safety, valve lever, and rocker arm in position in right valve grip. (Fig 48) Be sure the shorter end of rocker arm is at the top. Place grip spring over spring pin. Slip short end of grip spring in groove of rocker arm. Place long end of grip spring on outside of grip safety.

(2) Put left valve grip in place and insert the two lower lock washers and screws. Tighten the two screws enough to hold parts in place and still leave space for moving long end of spring into the groove in grip safety. Push spring into groove with a screw driver. (Fig 49)

(3) With spring in place, fully tighten the two screws with screw driver.

(4) Place grip support in position, and insert the two upper lock washers and screws. Tighten screws, using screw driver.

(5) Attach valve grip to valve body, using the four lock washers and inserting the four screws through the grip support. Make sure that the yoke shaft of the valve-diaphragm assembly is in front of rocker arm.

d. Maintenance of valve grip. No maintenance is required for the valve grip other than replacement of worn or damaged parts, tightening of screws, cleaning, and lubrication. (Par 49)

75. BARREL AND VALVE-BODY ASSEMBLY.

a. Description and functioning. (Fig 47) This assembly is part of the fuel valve. It includes the barrel, valve body, and operating parts contained in the barrel and valve body. The assembly consists of:

(1) Valve body, an aluminum housing, located at the rear of the gun and mounted on the grip support by means of four screws and lock washers. The valve body has four large threaded openings. The lower opening leads into the valve grip. The side opening, which forms a Y with the main portion of the body, is connected to the fuel-hose assembly. The front opening is screwed on the barrel. The rear opening is closed by the spring retainer and plug.

(2) Valve-diaphragm assembly, which transmits and reverses the movement imparted to it by the rocker arm of the valve grip. (Par 74 a) It also serves as a seal, keeping fuel from entering the valve grip. The valve-diaphragm assembly includes:

(a) Yoke shaft, on which the rocker arm bears at the lower end of the shaft.

(b) Yoke, a Y-shaped metal part which fits on the upper end of the yoke shaft and is held to it by a steel pin. The yoke transmits motion from the shaft to the yoke block, and is located within the valve body when the valve is assembled.

(c) Diaphragm, a synthetic-rubber diaphragm held in a steel sleeve, which fits snugly in the lower opening of the valve body. The yoke shaft passes through the diaphragm.

(3) Diaphragm support, washer, and cap, which hold the valve-diaphragm assembly in place in the valve body.

(4) Spring retainer, a brass, hollow bushing which screws into the rear opening of the valve body, and which is threaded internally to receive the plug. The retainer has a hexagonal head to take a 1-3/8-inch wrench. As its name implies, the retainer holds the valve spring in position.

(5) Plug, a brass part, resembling a cap screw, which fits into the spring retainer, closing off the rear end of the gun. It permits adjustment of the needle (see d below) without removing the valve spring and spring retainer.

(6) Valve spring, a coil spring located in the valve body between the spring retainer and the yoke block. The spring keeps the needle seated in the nozzle until compression of the grip safety and valve lever forces back the yoke block, spring, and needle.

(7) Yoke block, a steel piece, 1 inch long, which fits into the arms of the yoke Y. It is secured by an internal thread to the valve needle. Movement of the yoke in turn moves the yoke block and the valve needle.

(8) Lock nut, on the valve-needle thread at the rear of the yoke block, which locks the block on the needle.

(9) Valve-needle, a pointed rod, which extends through the inside of the barrel from the yoke block to the nozzle. The valve needle is seated in the nozzle except when firing. It controls the

ejection of fuel from the nozzle. Two sets of three fins each, known as needle guides, are mounted on the front and rear of the needle, respectively. These guides keep the needle centered in the barrel. The rear end of the valve needle is threaded to hold the yoke block and permit adjustment of the needle by means of the lock nut which screws on the threads. (See d below.)

(10) Barrel (Figs 47 and 54), which carries the fuel to the ignition head. It also supports or contains other components of the gun. The barrel assembly is replaced as a unit with the needle. It consists of a tube, made of thin metal, with a threaded fitting at the back end, and a nozzle brazed into the front end of the tube. The nozzle ejects the fuel from the barrel through the ignition head. The fuel emerges from two holes in the nozzle:

(a) Atomizer hole, a small opening which sprays a fine, readily ignited mist of fuel. This helps ignite the main stream of fuel.

(b) Main hole, which is tapered inside, and which conveys the main stream of fuel from the barrel. When the gun is not being fired, the valve needle is seated in the main hole of the nozzle. When the gun is being fired, the needle is withdrawn from the nozzle seat, permitting the fuel to be forced from the gun.

b. Removal of barrel and valve-body assembly. If gun group and tank group are connected, release any pressure in the fuel tanks by compressing the valve lever and the grip safety. Then disassemble as follows:

(1) Unscrew the fuel-hose assembly from the fuel-valve body only if this is necessary for maintenance.

(2) Remove spring retainer and plug from end of fuel-valve body and remove valve spring.

(3) Unscrew diaphragm cap and pull out washer, support, and valve-diaphragm assembly. To prevent loss of valve-needle ad-

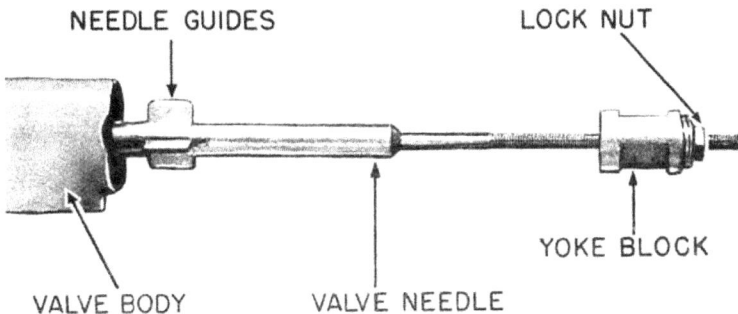

Fig 50. Valve needle, yoke block, and lock nut ready for installation in fuel-valve body.

Fig 51. Placing diaphragm
assembly in position in
fuel-valve body.

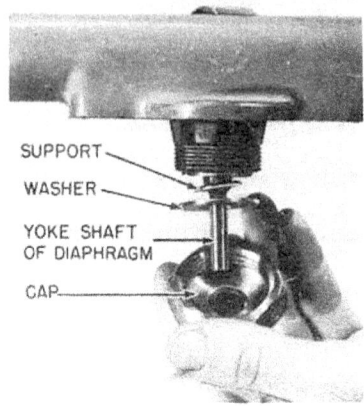

Fig 52. Installing parts
in fuel-valve body.

justment (Fig 54), do not disturb position of yoke block by turning the needle.

(4) Slide the valve needle out of barrel; the yoke block and the lock nut may then be unscrewed from the valve needle, but adjustment (see d below) will be necessary when reinstalling.

c. Installation of barrel and valve-body assembly.

(1) To install valve needle, screw the yoke block and lock nut on the needle (Fig 50). Insert needle in valve body and barrel.

(2) Insert valve-diaphragm assembly into valve body (Fig 51), making sure that the yoke slips into the flat notches of yoke block.

(3) Slip the diaphragm support, washer, and cap over the yoke

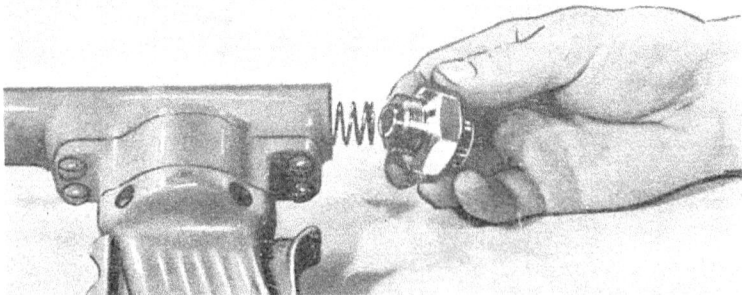

Fig 53. Installing spring retainer in fuel-valve body.

shaft. (Fig 52) Screw on the diaphragm cap by hand. Do not use a wrench. Install valve grip. (Par 74 c)

(4) Place valve spring over end of needle and install spring retainer. (Fig 53) Apply wrench very lightly to tighten spring retainer.

(5) Adjust needle (see d below), and screw plug into the spring retainer.

(6) If hose has been removed, apply anti-seize compound lightly to the threads. Screw hose into fuel-valve body. Wrench should be applied very lightly to tighten.

d. Adjustment of valve needle. Needle is adjusted after installation of parts in barrel and valve assembly. Use care when re-setting needle, as smooth operation of the weapon depends on accurate adjustment.

(1) Remove ignition shield (Par 18) and plug from gun.

(2) Use the valve-adjusting wrench (Fig 8) to hold the lock nut and apply a cabinet (narrow-bladed) screw driver (Fig 8) in the end of the needle. Turn needle until it makes a snug fit in the nozzle opening.

(3) Compress the valve lever and grip safety. The needle should draw back into the nozzle with the tip of the needle at the smallest diameter opening in the nozzle. (Fig 54)

(4) When the needle has been correctly adjusted, as in (3) above, tighten the lock nut with the valve-adjusting wrench, keeping the needle from turning with the screw driver. This will lock the adjustment. Screw plug into the spring retainer.

(5) Replace ignition shield. (Par 18)

e. Maintenance of barrel and valve body. (1) Damaged parts. Replace worn or damaged parts. If the diaphragm shows evidence

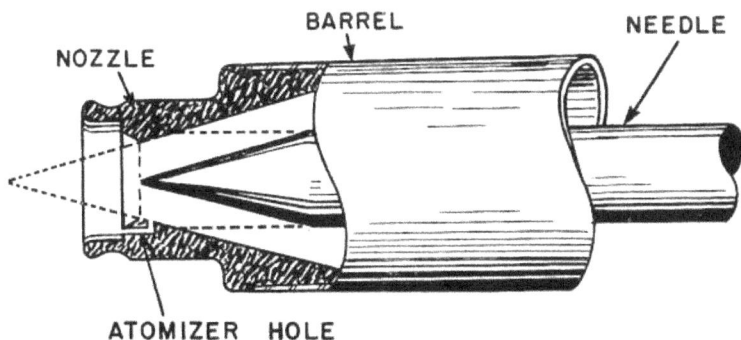

Fig 54. Valve-needle adjustment. Solid lines show needle in correct open position with point at smallest diameter of nozzle. Broken lines show needle in closed position.

of tears or separation, or if leaks occur at the diaphragm, replace the valve-diaphragm assembly.

(2) Valve spring. If valve spring has lost resiliency, grasp it by the ends and stretch slightly, or replace.

(3) Nozzle leaks. If valve leaks at nozzle, and cleaning (Par 55 d) does not remedy the leak, adjust needle (see d above). If leak persists, either replace barrel and needle, or lap seat. To lap, place lapping compound on seat (in nozzle) and on needle point. Turn needle in seat until parts make a tight connection when seated. Remove lapping compound, reassemble, adjust needle, and test fire.

(4) Atomizer hole. If atomizer hole is clogged, clean with fine wire. (Par 52 i)

76. IGNITION HEAD.

a. Description and functioning. (Fig 55) The ignition head ignites the fuel when the flame thrower is fired. It is mounted on the fore part of the barrel. It consists of:

(1) Ignition-head body, which includes half of the front grip. Three set screws serve to tighten the ignition-head body to the barrel. The ignition-head body is made of aluminum.

(2) Trigger and trigger bearing, held between the ignition-head body and the coverplate by the trigger screw.

(3) Trigger rod, one end of which is held in the trigger bearing, the other extending through the ignition-head body. Pulling the trigger shoves the trigger rod forward, causing it to push a match in the ignition cylinder. The match ignites an incendiary charge in the ignition cylinder.

(4) Trigger spring, which hooks over a projection of the trigger and is held at its lower end by a screw, which is held in the ignition-head body. This spring pulls the trigger rod back from the firing position after the firer releases the trigger.

(5) Latch, located in the ignition-head body, in front of and above the trigger guard. The latch, set on a pin, engages the notch of the ignition shield, locking it in place. A latch spring holds latch in position.

(6) Coverplate, an aluminum casting which constitutes the left section of the front grip and covers the working parts seated in the ignition head body. The coverplate and body are held together by four screws and four lock washers.

(7) Spring case, which turns the ignition cylinder when the trigger is pulled.

(a) Four projections on the inner spring case are bent over the outer spring case to hold the two parts together.

(b) The inner-case pin (Fig 56) engages a stop on the inside of the ignition cylinder. The five projecting metal matches on the inside of the ignition cylinder are each in turn stopped by the lug on the forward-facing surface of the ignition-head body.

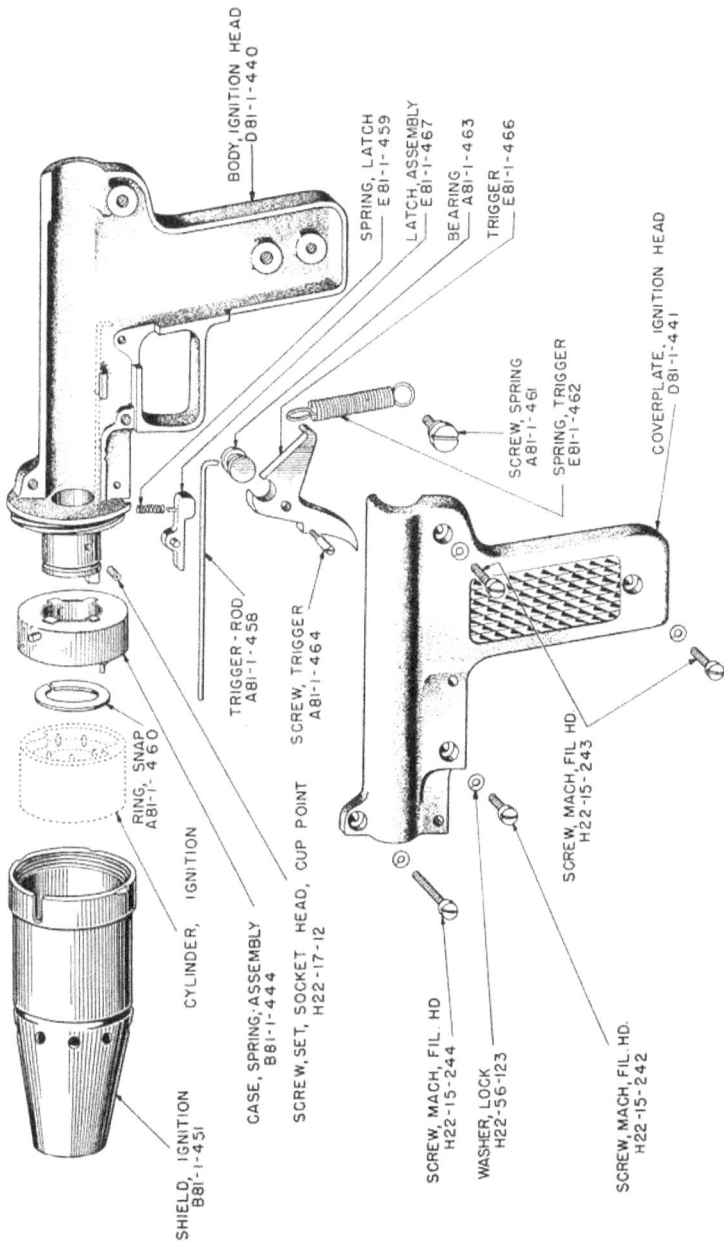

Fig 55. Ignition head disassembled, showing nomenclature and Chemical Warfare Service stock numbers for requisitioning spare parts.

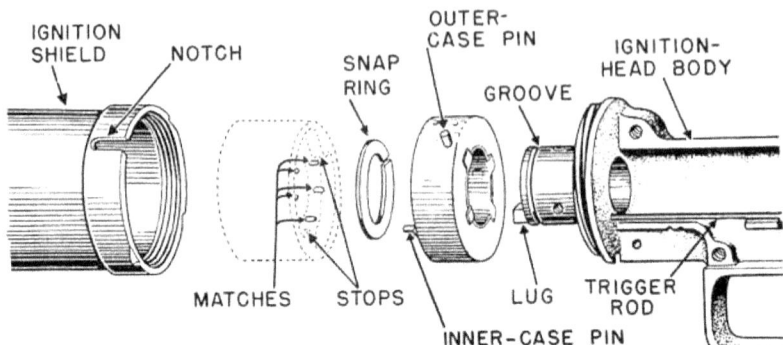

Fig 56. Parts of ignition head and ignition cylinder.

When the trigger is pulled, the trigger rod pushes a match forward, causing an incendiary charge in the ignition cylinder to
ignite. The spring in the case rotates the ignition cylinder until
another match is stopped by the lug.

(c) The outer-case pin (on the outside surface of the outer
spring case) fits into the notch in the ignition shield and holds the
spring case as the shield is screwed into position. This action
winds the spring in the case.

(d) A snap ring holds the spring case on the ignition-head
body.

(8) Ignition shield, a cylindrical, thin-metal tube with a con-
ical front end. The shield guides the flame and protects the firer.
Eight holes around the base of the cone provide an air intake for
burning the fuel. The base of the shield is threaded, and it screws
onto the ignition-head body. A notch (Fig 56) in the base of the
shield receives the latch and the outer-case pin of the spring case.

b. Removal of ignition head. To remove the ignition head, proceed as follows:

(1) Remove shield by lifting latch and unscrewing shield
counterclockwise. (Fig 14) Keep hands and face away from front
of barrel.

(2) If ignition cylinder has not been removed, remove it or
allow it to fall off barrel.

(3) Pry off snap ring which holds spring case in position, using
screw driver. (Fig 57) Be careful not to damage or break the
ignition-head body by applying too much leverage.

(4) Remove the four screws and lock washers which hold ignition-head body and coverplate together. Lift off coverplate.

(5) Trigger, trigger spring, trigger rod, latch, and latch spring
may be removed.

(6) Using a hex wrench, loosen set screws (Fig 58) and with-
draw barrel from ignition head.

c. Installation of ignition head. To install ignition head, proceed as follows:

(1) Insert barrel in ignition-head body, pushing it as far forward as the shoulder on the barrel permits.

(2) Aline front grip and valve grip.

(3) Using hex wrench, tighten set screws on barrel enough to hold but not so tight that barrel is dented.

(4) Place latch, latch spring, trigger and bearing, trigger rod, and trigger spring in position.

(5) Put coverplate on ignition-head body and replace the four lock washers and screws.

(6) Slip spring case over barrel, and lock by forcing snap ring into the groove.

(7) When weapon is to be used on a mission, fit ignition cylinder and ignition shield in place on nozzle end of barrel as described in Paragraph 18.

d. Maintenance of ignition head. (1) Servicing. The ignition head should be cleaned and lubricated each time it is disassembled. (Par 49)

(2) Spring-case assembly. If outer case rotates and inner case does not, and no spring action occurs, spring is broken and spring case should be replaced as a unit. Do not disassemble or repair this part.

(3) Trigger rod and lug. When trigger is pulled all the way, end of trigger rod should extend 1/16 inch beyond lug on forward-facing surface of ignition-head body. If end of trigger rod is worn, replace rod. Lug on the ignition-head body should be approximately 7/32 inch high. If lug is worn or broken, replace ignition head body.

Fig 57. Prying snap ring from ignition head to remove spring case.

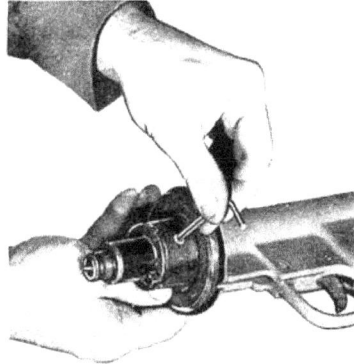

Fig 58. Loosening set screws with wrench so ignition head may be be lifted off barrel.

APPENDIX

Section XVIII SHIPMENT AND STORAGE

77. SHIPMENT AND STORAGE.

The flame thrower is shipped and stored in a wooden packing case (Fig 59), which measures approximately 34 inches by 23 inches by 19 inches. Cubage of the case is approximately 8-1/2 cubic feet.

a. Storage procedure. After use and servicing (Pars 55 and 56), if the weapon is not to be promptly reused on another mission, it should be returned to the packing case. Before disconnecting the gun group from the tank group and storing the weapon, the ignition cylinder should be removed, the fuel discharged, and the pressure released. Operate fuel valve to release any residual pressure in the fuel tanks. The deflector tube must be removed from the safety head (Fig 11) to permit the tank group to fit into the chest. The deflector tube should be kept in the spare parts

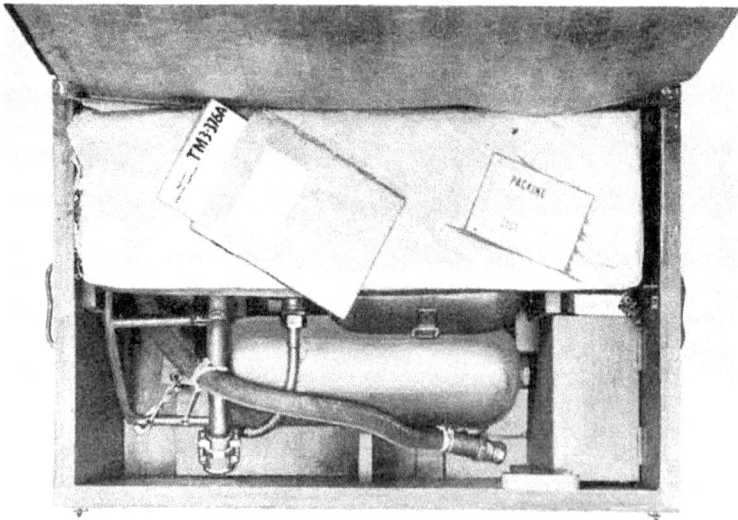

Fig 59. Opened packing chest showing flame thrower and other contents as received.

kit or tool kit until the next use of the weapon. The spare parts kit, the tool kit, the extra cans of cylinders, TM 3-376A, and the coupling plug (Fig 7) should remain in the chest except when they are being used. Wooden fittings hold the tank group in place, with the pressure tank up. The gun group is disconnected from the tank group and is kept with fuel hose connected to the gun on the gun mounting board in the chest. (Fig 10)

   b. Rust prevention. If the flame thrower, parts, and tools are to be stored for a considerable length of time, especially in a damp climate, all exposed metal surfaces should be covered with a rust-preventive compound. Store in a dry place.

Section XIX LIST OF REFERENCES

78. REFERENCES.

References pertaining to the care and use of flame throwers include:

AR 850-20	Precautions in Handling Gasoline
AR 850-60	Compressed Gas Cylinders; Safe Handling, Storing, Shipping, Using
FM 31-50	Attack on a Fortified Position and Combat in Towns
FM 100-5	Operations
TM 3-220	Decontamination
TM 9-850	Cleaning, Preserving, Lubricating, and Welding Materials and Similar Items Issued by the Ordnance Department

INDEX

- C Contd -

- D -

- E -

- F -

- F Contd -

- P Contd -

- S Contd -

- T -

- T Contd -

- V Contd -

NOTES

NOTES

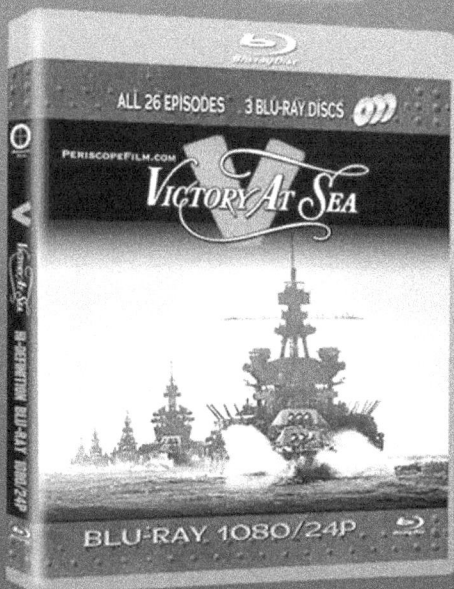

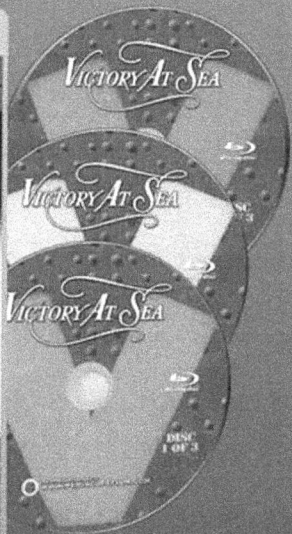

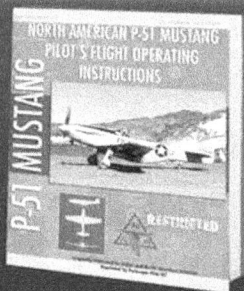

www.ingramcontent.com/pod-product-compliance
Lightning Source LLC
Chambersburg PA
CBHW052120090426
42741CB00009B/1885